Petroleum Engineering and Development Studies

Volume 3

Quality Assurance in the Offshore Oil and Gas Industry

PETROLEUM ENGINEERING AND
DEVELOPMENT STUDIES

Advisory Editor John S. Archer

Volume 1	Fundamentals of Casing Design Hussain Rabia
Volume 2	Directional Drilling T. A. Inglis
Volume 3	Quality Assurance in the Offshore Oil and Gas Industry J. H. Rogerson (editor)

Petroleum Engineering and Development Studies

Volume 3

Quality Assurance in the Offshore Oil and Gas Industry

edited by
J. H. Rogerson

Graham & Trotman
A member of the Kluwer Academic Publishers Group
LONDON/DORDRECHT/BOSTON

First published in 1988 by

Graham & Trotman Limited
Sterling House
66 Wilton Road
London SW1V 1DE
UK

Graham & Trotman Inc.
Kluwer Academic Publishers Group
101 Philip Drive
Assinippi Park
Norwell, MA 02061
USA

British Library Cataloguing in Publication Data

Quality assurance in the offshore oil and gas industry.—(Petroleum engineering and development studies; v. 3).
1. Offshore natural gas & petroleum industries. Quality assurance
I. Rogerson, J. H. II. Series
665.5′068′5

ISBN 0 86010 947 X (Vol 3, hardback)
ISBN 0 86010 932 1 (series)

LCCCN 87-12110

Printed in Great Britain at the Alden Press, Oxford

CONTRIBUTORS

K. P. Bentley
172 Hills Road
Cambridge
CB2 2BQ
United Kingdom

J. G. Hicks
35 Boxworth Road
Elsworth
Cambridge
CB3 8JQ
United Kingdom

J. H. Rogerson
Professor of Quality Systems
School of Industrial Science
Cranfield Institute of Technology
Cranfield
Bedford
MK43 0AL
United Kingdom

E. M. Rooney
Quality Assurance Training Services
School of Industrial Science
Cranfield Institute of Technology
Cranfield
Bedford
MK43 0AL
United Kingdom

J. Weston
Edward J. Weston & Company (Management) Limited
Level 2
Malt House
Coln-St Aldwyns
Gloucester
United Kingdom

CONTENTS

CHAPTER 1

INTRODUCTION

J. H. Rogerson

AIM OF THE BOOK

Everyone who is concerned with the offshore oil and gas industry is likely to be affected in some way by the formal discipline of quality assurance (and its complementary discipline quality control). The degree of involvement with the discipline will quite naturally vary and the importance attached to the various aspects will also vary. For example, a senior executive in an operating company will be most directly concerned with the cost effectiveness of quality assurance (does the expenditure of significant effort in devising and implementing a quality system lead to improved quality and greater reliability and efficiency?). On the other hand, a quality manager in a supplier company will be primarily concerned to ensure that his particular quality system is "acceptable" to his client. Again, an engineering specialist (a welding engineer or an inspection engineer perhaps) is mostly concerned to know to what degree of detail he needs to document, approve and control his procedures. These are all legitimate and necessary concerns and all form part of the overall quality system in an operation but the degree and extent of knowledge required of the formal discipline of quality assurance are obviously different in each case.

It is not possible to be all things to all men, so it is not sensible to try to write a book on quality assurance in the offshore oil and gas industry which will meet the needs of everyone; it is necessary to be selective. What we have tried to do is to take the major "issues" and deal with them in some depth, but at the same time in a practical manner. In this way we are trying to provide guidance in the application of formal quality assurance at a professional level. We are not presenting any arguments about whether or not quality assurance is a "good thing" or seeking to justify the use of formal quality systems. The industry has effectively settled this point by asking for formal quality assurance as a matter of course (and sometimes as a contractual requirement). What we are aiming to do is to discuss the philosophy of quality assurance in the context of the industry and show how this philosophy should be interpreted in terms of policies and procedures both in the range of policies and procedures needed and also in the degree of detail that is needed in each case.

Therefore, the aim of the book is to help in the definition, documentation and implementation of quality assurance systems. The book is firstly aimed at "quality" professionals in the industry (both in operating companies, contractors and suppliers) and secondly, and equally importantly, at *all* engineers and managers who need to define and implement technical procedures which must take due note of quality assurance requirements. In other words, design engineers, certification engineers, project managers, inspection engineers, welding engineers, in fact, most discipline engineers. The book is written at the level of the engineer/manager who needs to know about quality assurance. Perhaps not every engineer in the industry will need to fully understand every topic which is covered in this book but all engineers and managers will need to be conversant with some of the topics and therefore should find, at least some of the book, useful to read and to act as a reference in the interpretation of formal quality assurance criteria.

WHAT QA IS AND WHY IT IS IMPORTANT

QA is defined in a number of standards as will be described in chapter 2 but most of the "official" definitions—whilst not inaccurate—are not very helpful or informative. A more accurate definition is one which has been developed and used in Cranfield's courses for a number of years and is:

A *management system* designed to give maximum confidence that a *given, acceptable level of quality* is being achieved with *minimum total expenditure*.

This definition stresses the three major aspects of quality assurance—that it is a management activity (not just quality control inspection for example), that it assumes that the engineering requirements of the product have been adequately (and separately) defined and, finally, that the whole process is accomplished as efficiently and economically as possible. The definition also implies that "quality control" is a separate activity but one which is part of quality assurance. Quality control is, strictly, the physical actions of checking, inspecting and process monitoring which provide much of the evidence for the assurance of quality. We can say, therefore, that the management system of quality assurance is, in part, the management of quality control functions and so quality control is part of quality assurance. Inevitably, it is sometimes difficult to define whether a particular task is a quality assurance task or a quality control task but this is not really important (except to the pedants).

A most important distinction to be clear about, though, is the difference between the quality system and the engineering quality level of a product or item. As we have defined it, quality assurance is the means by which we

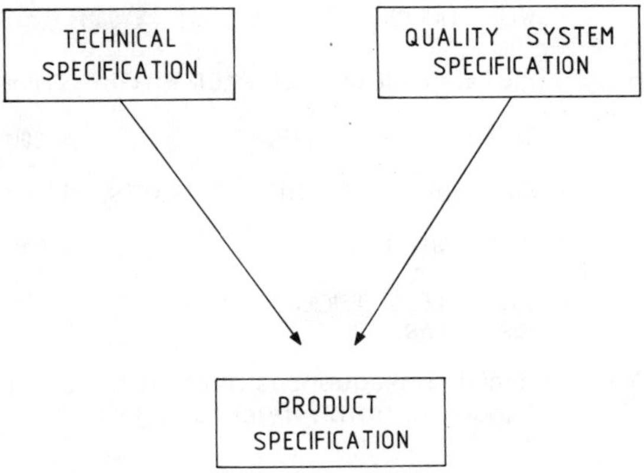

Fig. 1.1 Relationship between the quality specification and the quality system specification

manage the quality and, in chapter 2, we will discuss the methods and the standard criteria against which we assess a quality programme. The quality level (ie the performance characteristics, the tolerable level of defects or defective components, the precision of the design) is defined according to a technical standard or unique specification and is a separate (though related) activity. Figure 1 indicates this relationship. A product specification should include *both* the technical requirements and the requirements for assuring that these technical requirements are complied with. It is this latter subject which we call quality assurance.

We talk about technical specification and quality specification. These are not necessarily the same things as *standards*. A standard is a set of general requirements which are applicable to a wide range of circumstances and which are widely acknowledged as being appropriate—in other words a standard is a consensus document. The technical, or quality system requirements, for a project may be so specific (and almost certainly will be in the offshore industry) that a consensus document will not be precise or detailed enough. Consequently, standards will need "interpretation" or amendment for particular circumstances. This will be discussed in more detail in chapter 2 when the quality system standards are described.

Quality assurance, as a formal discipline, was developed in the "project" type industries where clients purchase expensive and complex capital plant. The philosophy naturally percolated into other industries and is now widely accepted as a necessary discipline for *all* types of manufacture. Nevertheless, the primary impetus came from the clients in the capital plant sectors (Defence Ministries, Public Utilities) and the philosophy of the formal discipline of quality assurance was established to meet their needs (not necessarily the needs of their suppliers). It is a fact that it is the client

	POWER STATION PROJECTED COST	$700M
IF A 1 YEAR DELAY ON ONE CONTRACT WORTH		$100M
THEN	1. EXTRA INTEREST CHARGES.	$ 80M
	2. ADJUSTMENTS TO OTHER CONTRACTS.	$ 40M
	3. EXTRA SITE COSTS.	$ 20M
	4. USE OF LESS EFFICIENT PLANT FOR 1 YEAR.	$ 60M-200M

Fig. 1.2 Financial consequences of construction delays (derived from reference 1.1)

who is most exposed to the consequences of quality failures both in terms of the safety hazards (fire, explosion, release of toxic gases, release of radio-activity) and in terms of the economic effects (late delivery leading to commissioning delays, unplanned shutdowns of process plant). Figure 1.2, derived from data presented by the Central Electricity Generating Board in the UK in the late 1970s (ref 1.1), indicates the magnitude of the financial consequences of a major quality failure (in this case, delay in delivery of a major subsystem) and the evidence of major disasters (Flixborough chemical explosion, Sea Gem collapse and the structural collapse of The King's Bridge, Melbourne, being well documented examples) indicates the safety implication of a failure of quality. Consequently, it is not surprising that the formal discipline of quality assurance was developed by these "project" type industries as the *only* means of guarding against the risk of major and costly quality failures. Quality Assurance is the "insurance premium" that the client pays to guarantee (as far as is reasonable) the quality of what he purchases. The content, style and emphasis of the discipline tend, therefore, to reflect this client concern. To such a client, formal quality assurance is a necessary part of a project and can be seen as an investment in the same way as proper project planning is an investment—a proper use of engineering and managerial resources to attain the satisfactory completion of the overall task.

SPECIAL FEATURES OF THE INDUSTRY WHICH RELATE TO QUALITY POLICIES

Much of the available literature on quality assurance and quality control is written from the point of view of "conventional" engineering manufacture. Also, as we shall see in the next chapter, the wording of the major quality system standards is primarily relevant (or is most easily interpreted) for such industries.

The offshore oil and gas industry has, however, some unique features

OPERATORS	Avoidance of large consequential costs.
CONTRACTORS	Guaranteeing that operators stated performance requirements are complied with.
MANUFACTURERS	Meeting specification with minimum cost.
INSPECTORATES	Assurance that items are safe and meet defined quality standards.

Fig. 1.3 QA objectives of the different participants in a project

(which are shared to a degree by other energy industries) which influence very strongly the philosophy and practice of quality assurance. The most important characteristic of the industry is that the end product—the offshore installation—is of very high value. It is of high value in two respects. First, obviously, the cost of the design, manufacture and installation is high; therefore, the capital investment in the plant is high. Secondly, and perhaps less obviously, the high investment in the plant is only justifiable if it is operating continuously and at a high level of capacity, ie unplanned outages are expensive.

From the point of view of the operator, therefore, (the organisation which paid for the plant and stands to lose if it is not operating) assurance about the quality of the installation is at least as important as assurance about project and operating costs and project completion times. The subject of the cost implications of quality assurance programmes will be discussed in detail in chapter 7 but the point to recognise here is that the operator has a very direct interest in the assurance of quality so he will inevitably require his suppliers to devise and implement quality systems to meet his (the operator's) needs and not *necessarily* the supplier's needs. Although, of course, all parties have a vested interest in quality and its assurance, the emphasis is different and the cost/benefit trade offs are different between the operator, supplier, contractor and third party inspectorate or regulatory body. A regulatory body or inspectorate will certainly have a vested interest in the assurance and control of quality but with the prime aim of being assured that the structure is safe and complies with the appropriate regulations (hopefully the same thing!). Such a body is not *directly* concerned with the cost/benefit aspects of quality assurance. Figure 1.3 shows, in a simple way, this difference in attitude between the major parties. Each of these attitudes is acceptable but what any formal quality system must do is reflect these different attitudes (which relate, really, to the degree and type of risk being incurred) and preserve a realistic balance between them. This is, clearly, not always an easy task. This is the cause of some of the conflicts of interest in the application of quality systems in practice. Remember, it is a "buyer's market" in that the operator decides what assurance of quality he requires because it is he who stands to lose most if quality is inadequate.

Another feature of the industry is that the end product is unique. We are not talking about a mass production industry where a supplier manufactures "standard" products to his own specification. Every project will be different and, in most cases, suppliers will be working to a very precise specification issued either by, or on behalf of, the operator. The project may well involve the supply of "standard" items—structural steel, pumps, valves, instrument panels, for example, but these will usually be supplied as part of a manufactured item or system. This uniqueness of the supply means that suppliers will need to develop quality procedures and even quality systems which are unique to the project—ie there is a need for "project specific" quality systems. The supplier often does not have the opportunity to streamline his quality system in the way that a manufacturer of mass produced systems can, so that the cost (to the supplier) of applying formal quality systems can be high. Also the client (or his contractor) in effect, will dictate the quality system approach to be used. This goes against the traditional view of quality assurance in which the customer, by implication, defines certain quality system criteria and leaves the supplier to meet these criteria in his own way. This recognises the fact that each supplier is different. He has a different management structure, different facilities and a different way of doing things which it is not sensible for a customer to interfere with. In the offshore industry, on the other hand, the fact that the customer defines the requirements so precisely means that he needs to define, in some detail, how the requirements are to be met—what inspection requirements, what sequence of manufacturing operations, what documentation requirements, for example. The interface between the supplier's organisation and the client's organisation must be well controlled. In fact the large and complex nature of many projects means that there are a number of interfaces to be controlled between operators, contractors, subcontractors, raw material and component suppliers and certification bodies. The contractual chain can be complex and, as was stated above, not all the participants have the same priorities and aims in respect of a quality programme. Figure 1.4 shows an example of a typical contractual chain for the supply of one item (a module) which illustrates the interface problem (another example is shown in chapter 3). In this particular example the communication difficulties across these interfaces and the consequent lack of commonality in the definition of quality levels and how far they should be assured led to the quality related costs of the project being excessively high and to the imposition (with hindsight) of some unnecessary demands on the supplier (ref 1.2).

Another aspect of this uniqueness of the product (and its high value) is that the product when delivered or installed must conform to every requirement of the client's specification unless a concession or variation to contract is negotiated. In a "quality" sense this means that the supplier must always aim to supply "defect free" product. We could argue that

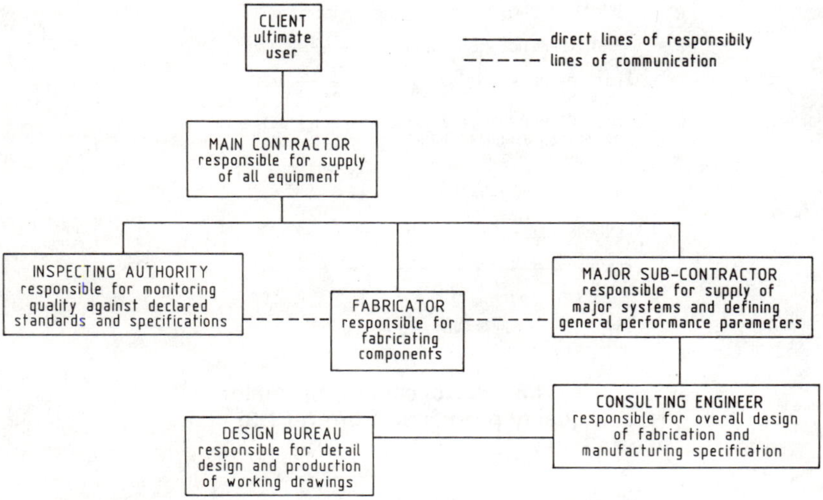

Fig. 1.4 Example of a contractual pattern for fabrication (ref 1.2)

suppliers of any product should aim to supply defect free product. However, in the case of mass produced items of relatively low intrinsic value, it is sometimes unreasonable and uneconomic to do this. It may be, in such cases, better to rely on a final inspection system which prevents defective product being shipped or even to budget for a small number of warranty claims from customers. Clearly, such policies would not be appropriate for suppliers to offshore oil and gas projects so they must have quality systems which are good at *preventing* failures or defects arising.

The offshore industry, particularly in recent years where ever more hostile environments and deeper waters are being encountered and where the need for novel and cheaper methods of producing from marginal fields and where the need to extract, economically, a greater proportion of the available oil and gas from existing fields have arisen, is very much a "high technology" industry in the areas of design of installations, manufacturing methods and materials used and in the process control and safety systems used in operation. The introduction of novel technology is, in itself, a major quality risk. A survey some years ago by the British Institute of Management showed that the use of new technology in terms of design, materials and manufacturing methods was an important root cause of major quality problems. Figure 1.5, developed from the results of this survey, goes further and indicates that "pre-manufacturing activities" are the source of at least half the major quality problems in engineering manufacture. Note this survey relates to the generality of engineering manufacture. We can assume, therefore, that for "high technology" industries these pre-manufacturing activities will be of even greater significance as the proportion of "new technology" will be higher. This implies that a quality system must pay

HUMAN ERROR	12%	
BAD INSPECTION METHODS	10%	
LACK OF OR WRONG SPECIFICATION	16%	52%
LACK OF PROVING (NEW DESIGNS, MATERIALS, MANUFACTURING METHODS)	36%	
POOR PLANNING	14%	
UNFORSEEABLE	8%	
OTHERS	4%	

Note:- Pre-manufacturing errors caused 52% of problems.

Fig. 1.5 Basic causes of major quality problems (from ref 1.3)

particular attention to the control of design and development work, planning and procurement. The "traditional" attitude to quality assurance has been characterised by much emphasis on manufacturing and inspection. These activities are, of course, important to the quality of a supplied item but, in the offshore industry, the emphasis of quality assurance should swing more to the pre-manufacturing stages. If these are not controlled properly then control of manufacture is almost a waste of time.

The industry is also special in that, in many cases, (particularly in the North Sea) there are regulations which must be complied with before an installation is permitted to operate. The rules vary in different countries but the concept is the same. There must be a "certification process" which demonstrates that the installation is fit and safe to use. "Certification" will affect all stages of an offshore project from design through to commissioning and has, historically, been seen as an additional (and costly) extra engineering activity. It would seem obvious that a properly constituted quality programme should provide the evidence (or most of the evidence) that an installation is safe and fit to use. Therefore, there should be a close interface between the quality assurance and the certification activities at all stages. From the operator's point of view, it might be appropriate to use the certification requirements as the basis for defining the details of the quality programme. If you implement a formal quality system and it does *not* cover the certification requirements then it is an unnecessary duplication of effort.

As a corollary to this need for initial "certification", it is a requirement that an offshore installation maintains its fitness for use over a period of many years.

This means the structural state of the installation, the process and other operating equipment, the support services and the safety systems need to be maintained and kept to a defined level of capability. A formal quality system to cover these activities is becoming a necessity. Of course, other types of installation (land based oil and chemical installations, power generation equipment for example) also have this requirement but the particularly hazardous nature of many offshore operations and their physical isolation means the requirement for the assurance of quality of the operations becomes more significant—not least as a means of efficiently meeting the regulatory requirements.

Finally, a feature of the industry which is relevant to the quality assurance approach is the very great importance of the timely completion of projects. Offshore projects are expensive and they are financially justified by the fact that they can generate large cash flows when operating. A delay in operation is very serious so there is *always* pressure to keep to project timetables. This may introduce awkward dilemmas when the assurance of quality is concerned, ie there is a temptation to take short cuts. The implication for the quality system approach is clear—it must be an efficient and minimum programme in the sense that unnecessary actions are not required and the emphasis must be on the planning and other pre-manufacturing stages to ensure, as far as is possible, that project quality is designed and planned in.

IMPLICATIONS OF THESE SPECIAL FEATURES OF THE INDUSTRY

As a consequence of these special features which have a bearing on the approach to quality assurance, we can define certain key aspects of a quality system which are necessary for it to be effective. These aspects must be properly addressed and, furthermore, must be very much borne in mind when applying the "general" quality system criteria defined in the various standards for quality systems (see chapter 2). This is not to say that these quality system standards are inappropriate—merely that they are designed to be applied very widely to the supply of *all* types of products and services (not just engineering products). As a consequence of this, they need to be "interpreted" in the context of the applicable industry (such that some of the criteria are less relevant than others, for example).

We can, therefore, summarise the special aspects of the quality system requirements for the offshore oil and gas industry as follows:

1. Control of Interfaces—the complexity of the projects and in consequence the large number of parties concerned who will not all share the same priorities in terms of the details of the overall quality programme means communication, coordination and correlation between all the participants in a total quality system is crucial.

2. Control of Design and Procurement—the "high technology" nature of the industry with its significant use of novel technology

together with the financial (and other) costs of failure and delay in installation mean that the pre-manufacturing activities (the "front end" engineering and project costs) are of *relatively* greater importance than the manufacturing phases.

3. Control of Certification—the regulatory requirements require the demonstration that an installation is safe and fit for use so the quality programme must encompass the requirements of the certifying body.

4. Control of Operations and Maintenance—the hazardous nature of many of the operations and the complexity of the facilities, the economic consequences of unplanned shutdowns and the need to maintain certification require the application of the quality system approach during the operational phase.

5. Control of Quality Related Costs—reasons can always be devised for increasing the rigour or extent of a quality programme especially when the potential consequences of quality failures are so serious and therefore quality programme costs can easily get out of control if not carefully monitored.

STRUCTURE OF THE BOOK

This book is structured, therefore, to stress the importance of these key features which are specific to the industry.

Apart from this introductory chapter, the book contains six chapters, each of which deals with one major topic.

Chapter 2 is concerned with the quality system standards. All quality systems must be designed to meet the criteria of one or other of the major quality system standards. These standards define, in very general terms, management criteria which must then be interpreted in the context of a particular industry, product or business. It is an unfortunate fact that standards are in most cases written for the generality of manufacturing and is not always immediately obvious how they should be applied to the offshore industry. Nevertheless, engineers and managers within the industry must be aware of these standards, the differences between them and the significance, in a practical sense, of the various criteria. Of particular concern is their application to operation and maintenance activities.

Chapter 3 takes the major subjects of project management, design and procurement, ie the front end activities. It is a fact, not, of course, unique to this industry, that decisions taken at this stage are fundamental with respect to the outcome of any project. Errors made at this stage can be corrected relatively painlessly and changes made which could be difficult, if not impossible, to make at later stages. The assurance and control of quality at these pre-manufacturing stages is therefore self-evidently important and is, arguably, the *most* important stage in the total quality scene.

Chapter 4 covers construction and installation. This subject is, in a sense, the most visible and most easily understood part of the quality activity. It deals with the quality of the hardware and it is related very closely to the traditional topics of inspection, approval of manufacturing procedures, production control and "hold" points. It is also the labour intensive part of QA and, therefore, the relatively costly part. It then becomes a matter of great concern that it is carried out as efficiently as possible. Communications and interface control are particular problems in this stage because of the potentially large number of suppliers and supplied items.

Chapter 5 looks at the role of formal quality systems in helping to satisfy certification requirements. Certification is, particularly in the North Sea environment, a statutory requirement for an operator to satisfy before he is able to use his facilities as it demonstrates that his installation is safe to operate. In the early projects, this was a costly, inspection based stage. Clearly, a good quality system should ease the path to certification and enable the certification activities to be more efficiently integrated into the project planning and execution. Certification in some form or other is likely to be needed in any offshore installation so the quality system instituted by an operator must take into account certification needs.

Chapter 6 is, in effect, an extension of Chapter 5 as it addresses the problems of operation and maintenance and, therefore, the maintenance of certification. Historically, formal quality systems have been restricted to design, manufacture and installation. Operation and maintenance have been rather ignored from the quality assurance point of view (this, again, is not just a feature of the offshore oil industry) but any logical analysis of the operation of an offshore facility indicates that formal quality systems are just as necessary as in the design and manufacturing stages. As well as the maintenance of the fixed structures, it is also necessary to consider the process equipment, the fire detection systems, the in-service inspection and monitoring, the diving operations, in fact, *all* operations which take place in connection with the production and transport of oil and gas.

Finally, chapter 7 deals with the cost and cost effectiveness of QA operations. The aim of investing resources in a formal quality system is to achieve more economically and more reliably the quality of installation (or operations) which is required than would be possible *without* a formal quality system. There is, inevitably, concern that a formal quality system should be cost effective and this is not always easy to measure, nor is cost effectiveness perceived in the same way by operators as by contractors or suppliers. Nevertheless, the subject is important and it is possible to identify and measure the costs of quality related activities and make some logical estimates of the required scope and detail of a quality system for a particular project.

REFERENCES

1.1 Peddie, R. A. *The management of large, high technology projects*. The Institution of Electrical Engineers. February 1976.
1.2 Grant, I. M., Rogerson, J. H. *The Importance of contractual requirements in determining quality costs in the fabrication industry*. W I Conference November 1981.
1.3 Belbin, R. M. *Quality calamities and their management implications*. British Institute of Management. Report OPN8 1970.

CHAPTER 2

QUALITY SYSTEM CRITERIA AND THE APPLICABILITY OF QUALITY SYSTEM STANDARDS

J. H. Rogerson

In chapter 1 we defined quality assurance and explained how a product specification should include both the technical specification (the performance requirements and the quality requirements) *and* the quality assurance specification (the extent of assurance that the technical requirements will be met). Our definition of quality assurance is, therefore, primarily a definition of a management system and a quality assurance specification must, in its turn, specify a management system. A specification for a management system is not as easy to visualise and, at first sight, not as easy to define as a technical specification. However, if you analyse any technical specification, you will see that it defines technical criteria which must be conformed to (eg strength or toughness criteria for a steel or sensitivity and reliability criteria for fire detection equipment). It is just the same with a specification for a management system which provides assurance of quality—the specification defines criteria which must be conformed to.

The reason why specifications of all types should define criteria is that it is not sensible, or usually possible, to define *how* an organisation does things (whether it be how it should make steel or how it should assure the quality of its products) because every organisation is different and will do things in different ways. What it *is* sensible to do is to define, as precisely as possible, what criteria (be they technical or managerial) an organisation should meet.

This specification of criteria (rather than specification of *how* to do something) is of fundamental importance, particularly in the case of a quality system specification. All the good quality system standards and specifications limit themselves to defining criteria. It is the task of the users of these specifications (supplier, purchaser and any other interested party) to interpret them and apply them in the context of the particular situation—this is where the skill lies in devising effective quality systems.

Defined Responsibilities.

Defined Procedures.

Defined Follow up Action.

Adequate Records.

Adequate Communications.

Fig. 2.1 The five features of a quality system

THE MAJOR ELEMENTS OF A QUALITY SYSTEM

A formal quality system, can in principle, be applied to any activity, although, in practice, formal quality systems are mostly only applied to the activities of design and manufacturing. This will change, however, and in the case of the offshore oil and gas industry operations (and later de-commissioning?) will need to be covered by a quality system approach.

Whatever activity is subject to a formal quality system, we can define five features of the management system which must be in place (and be worked to). These are shown in Figure 2.1. These five features are features which should be part of any management system but, in respect of quality assurance, they have a special significance.

"Defined responsiblities" means that all staff (not just the quality manager) should have their responsibilities and organisational relationships defined in respect of quality matters. Design, production, purchasing, contracts and inspection staff all have an influence on the assurance of quality in the offshore industry so their responsibilities in this area should be defined.

"Defined procedures" means that the various management tasks (eg documentation control, authorisation of design changes, selection of suppliers) as well as the technical tasks should be carried out according to defined rules of procedure. This is important for consistency accross an organisation and over a time period—particularly where personnel can change or an organisation is operating on different sites.

"Defined follow up action" is a recognition of the fact that every "action" generates the requirement for a response or some feedback. For instance an inspection generates an action which can be "accept the item"; "reject the item" or "repair the item". For the same reason that there should be defined procedures (consistency) there need to be defined follow up actions.

Follow up actions (which include feedback) are also the way of self-regulating a system and ensuring that quality is being maintained.

"Adequate records" are clearly necessary as the assurance of quality always includes the provision of evidence (ie records) to show that satisfactory quality is being achieved. A further reason for adequate records is that it is only on the basis of records that a quality system can be monitored to see that it is effective.

"Adequate communications" are perhaps the most important feature of all because unless communications within an organisation are adequate then defined responsibilities, procedures and follow up action are pointless. Any quality system is about people and relies on people to operate it. Communications are fundamental. Likewise, communications between the parties in a project are of fundamental importance. Most offshore projects are complex and involve many organisations working together. Communication on quality matters across the interfaces between different organisations needs emphasising in all project planning.

THE MAJOR OPERATIONAL ACTIVITIES

We have defined the major elements of a quality system; we must now define the major activities in an operational sense. Each of these major activities must be conducted in accordance with the management elements previously defined. Thus we can define a quality management system for the design and manufacturing phase of a project as a matrix of these elements and activities (see Figure 2.2). The construction of a formal quality system, therefore, means the construction of such a matrix and it is the task of a quality system specification to define the requirements of this matrix.

These major operational activities can be listed as follows and the flow charts in Figures 2.3 to 2.9 illustrate the stages of control.

1. Design

Conceptual design or the generation of the design concept is the only non-controllable activity. Other activities in design relate to the transmission of information, design calculation, checking and verification and design review. All of these must be carried out in a systematic, controlled way to *provide assurance* that the design has been carried out to the intent of the client's performance specification.

2. Procurement of Material and Parts

Technical requirements must be properly and fully defined and there must be *assurance* that this has been done. There must also be *assurance* that

	Design	Procurement	Manufacturing	Inspection & Test	Calibration & Control of Test Equipment	Corrective Action	Handling, Storage & Despatch
Defined Responsibilities							
Defined Procedures							
Defined Follow up Action							
Adequate Records							
Adequate Communications							

Fig. 2.2 Quality management matrix

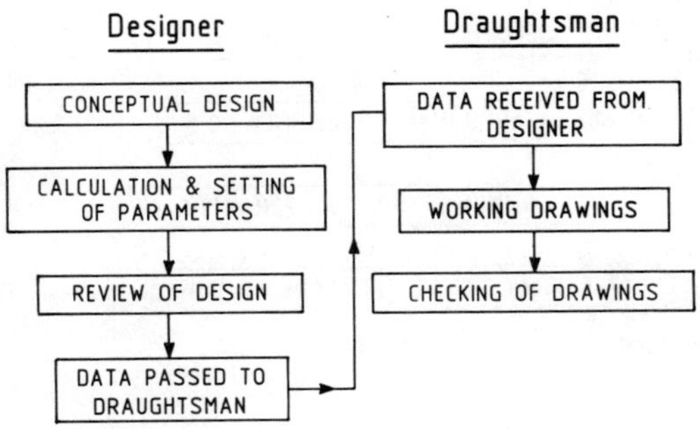

Fig. 2.3 The design process

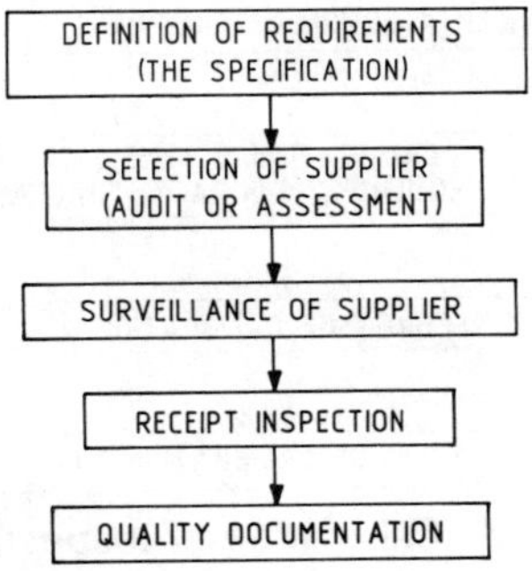

Fig. 2.4 Procurement

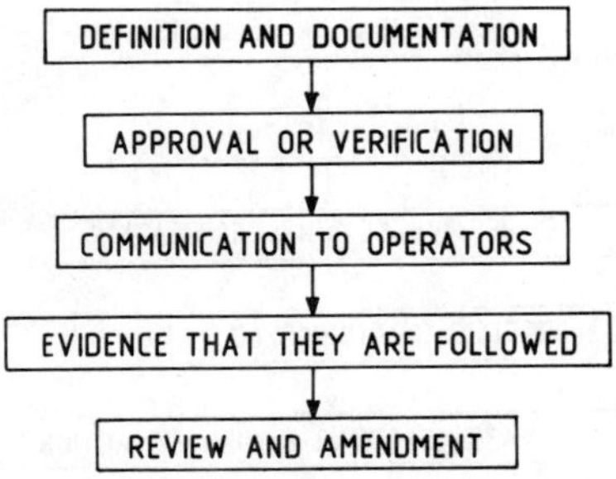

Fig. 2.5 Manufacturing procedures

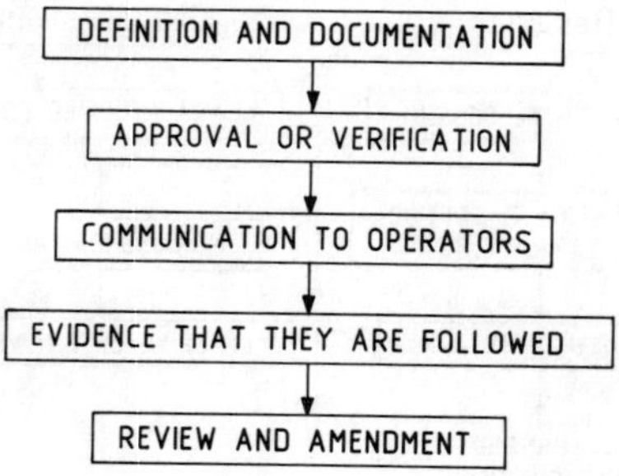

Fig. 2.6 Inspection and test procedures

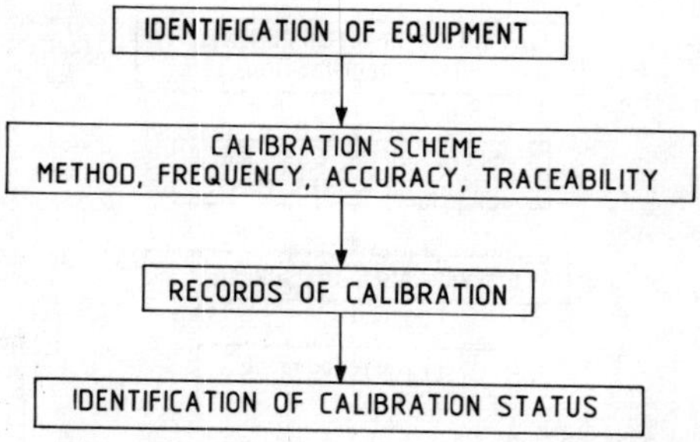

Fig. 2.7 Calibration of inspection equipment

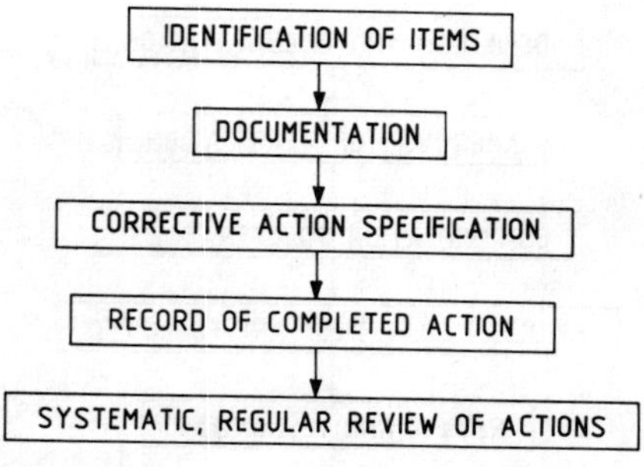

Fig. 2.8 Non-conformance and corrective action

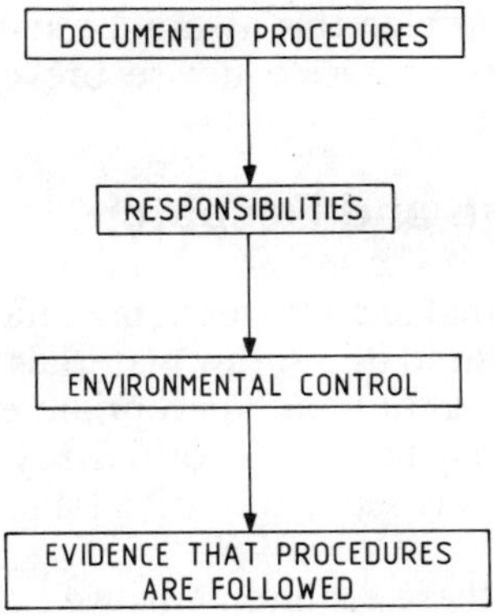

Fig. 2.9 Handling, storage and despatch

suppliers have been selected on the basis of the quality of their products and the adequacy of their quality systems. In any project purchased items form a large portion of the value of the total package.

3. Manufacturing

To provide *assurance* that all manufacturing is carried out in a correct manner, there must be manufacturing instructions and procedures which are correctly authorised, transmitted to the appropriate works staff and, above all, adhered to.

4. Inspection and Test

There must be assurance that adequate and suitable procedures have been defined and properly followed by appropriately qualified staff. The inspection and test results must, of course, be properly recorded and documented.

5. Calibration and Control of Inspection Equipment

All such equipment must be identifiable, controlled and subject to a defined calibration scheme which is properly adhered to and which is appropriate for the level of accuracy required and type of instrument. No inspection or test result is worth anything unless the inspection and test equipment can be shown to be in working order and within calibration.

6. Corrective Action and Non-conformance Control

There must be *assurance* that all defectives are identified, segregated in some way, and reworked, scrapped or repaired as appropriate. This must be documented both to provide this assurance and, in the longer term, to permit analysis of quality trends to devise preventive measures for the future.

7. Handling, Storage and Despatch

There must be *assurance* that at all stages of manufacture, storage, despatch and installation manufactured items, raw materials and semi-finished product are properly handled, kept in an appropriate environment and appropriately packaged. It is easy to damage or destroy expensive and valuable equipment by neglect of this seemingly trivial and obvious requirement.

Over and above each of these requirements we need a further element of the quality system and this is what is generally called "the system review". In effect, this is the basic monitoring operation to see that the quality system is understood, is being followed and is adequate to provide the correct degree of assurance. Such a monitoring activity must be carried out by means of internal audits of the system together with systematic and regular analyses of the quality performance (eg analyses of quality trends, areas of excessive quality related costs, continuing or "type" quality problems). Together, such monitoring operations are necessary to keep a quality system in a good condition and relevant to the tasks required of it.

The control of these activities includes, of course, the planning of operations in a formal sense and the accompanying documentation and records. In the major quality system standards (see a later section of this chapter) the control of these activities is defined under up to twenty different headings. However, the *major* activities which really determine the adequacy of a quality system can be considered under the seven headings listed above.

The control of operations and maintenance (and of the provision of services) is rather different and will be discussed in chapter 6. In fact, as we shall see, the emphasis, to date, on formal quality systems has been entirely on the design, manufacture and installation stages of projects. This is, of course, illogical but none the less true and the relevant quality system standards and specifications reflect this bias.

DEFINING THE QUALITY SYSTEM CRITERIA

To define the quality system criteria for an offshore project (or a portion of it)—ie to define the detailed matrix requirements—requires a careful, detailed specification taking due note of the interfaces. No one specification for a quality system will be *exactly* the same as another because of the varying nature of projects. However, there are some basic standards for providing the framework of a matrix. These we will now discuss.

This discussion is not going to be a detailed clause by clause analysis but a discussion of the differing emphasis and style of the various quality system standards which all look (at least superficially) the same. They are different, however, in some important details although they all address the same major features and operational activities. You need to read and study the standards themselves to be fully aware of the significance of the quality system criteria and how they differ from one standard to another. This discussion, therefore, should be looked upon as a commentary to be read in conjunction with the appropriate standards.

There are three basic types of non-military standard (ie standards which define the criteria for a formal quality system). These are:

1. The standards for nuclear safety related items:
   eg BS 5882 (UK)
   ANS1 N45.2 (USA)
   CSA Z299 level 1 (Canada)
2. The general purpose standards:
   eg BS 5750 parts 1, 2 and 3 (UK)
   NS 5801-3 (Norway)
   CSA Z299 levels 2, 3 and 4 (Canada)
   ISO 9000-9004 (International)
   SABS 1057 (South Africa)
3. The Industry Specific Standards
   eg API-Q1
   ASME Section viii

The different standards in group 1 are very similar (almost word for word the same in places) and the different standards in group 2 also have much similarity with each other although there are some differences between them. The group 3 standards are, of course, rather different but do bear some relation to each other and to the group 2 standards. It is important to remember that *all* quality system standards derive from the original military standards of the 1960s and this means that their layout, philosophy and, even, phraseology is very similar. However, it is the *differences* which you must be aware of.

The "nuclear" quality system standards are, in fact, of great potential interest to the offshore oil and gas industry because the two industries (nuclear and offshore) share many characteristics—high criticality, complex and novel technology, large and expensive projects. In fact, if we removed the word "nuclear" from the nuclear quality system standards we would have a very suitable set of criteria for the offshore oil and gas industry. On the other hand, the "general" quality system standards are of less direct value because their very generality makes them difficult to interpret in the context of a particular industry (particularly one with the specific characteristics of the offshore oil and gas industry). Nevertheless, their very existence as the official "non-nuclear" standards makes them important.

The industry specific standards—the third category of standards are different again. They truly address the specific quality needs of a particular sector of industry. For example, API-Q1 relates (at the moment) specifically to Wellhead equipment whilst the ASME Boiler and Pressure Vessel code Section viii relates specifically to fusion welded pressure vessels. This type of standard tends, naturally, to be more detailed than the more general standards, covers technical rather than just quality system activities and, frequently, indicates *how* quality activities should be carried out rather than just stating the criteria.

These basic differences between the three types of standard need to be understood because, although they all address the same subject, their approach is different. To explore these differences in more detail we can best start by discussing the "general purpose" standards and illustrate the difficulty in interpreting them in the context of the offshore oil and gas industry. We will then illustrate, by reference to both the "nuclear" and the "industry specific" standards, how more complete definitions of the quality system criteria can be made.

At the time of writing (1987) the new international general purpose standards (ISO 9000—ISO 9004) are just being issued and they will, it is hoped, be *the* established general purpose standards. They are, in fact, modelled quite closely on the British Standard (BS 5750) but with some modifications and improvements. In fact, the 1987 revision of BS 5750 will be made identical to the ISO standards.

However, for the moment, the ISO standards are not in general use and many organisations have naturally modelled their systems on the criteria of the existing national standards. A knowledge of these current standards is therefore necessary if only to identify their limitations and, hence, the limitation of quality systems which only meet the minimum criteria of these standards. This is a particularly important point in the context of the offshore oil and gas industry.

THE BRITISH STANDARD BS 5750 "QUALITY SYSTEMS"

This is a very typical "general purpose" standard in its approach and content which has set the general pattern for many others and, also, the new ISO standards. The only major standard which has a significantly different approach is the Canadian standard CSA Z299.

BS 5750 is in three parts, each of which has a different scope and range of applications:

Part 1: Specification for Design/Development, Production, Installation and Servicing

This part is to be applied for use "where a contract between two parties requires the demonstration of a supplier's capability to design and supply product".

Part 2: Specification for Production and Installation
This part is to be applied for use "where a contract between two parties requires the demonstration of a supplier's capability to control the processes that determine the acceptability of product supplied".

Part 3: Specification for Final Inspection and Test
This part is to be applied for use "where a contract between two parties requires demonstration of a supplier's capability to detect and control the disposition of any product nonconforming during final inspection and test".

From this definition of the scope of the different parts of the standard it is clear that part 1 is applicable to a situation where there is significant design work. This means, obviously, the major contractors and the suppliers of components or items which are specially designed for the project—in other words, part 1 is relevant to all the innovative work in an offshore project.

Part 2 is applicable where there is no design work involved. This means suppliers who are manufacturing to established designs or to designs supplied to them by the major contractor or the operator. Clearly, in this case, a less rigorous quality system is needed because the design process (the source of approximately half the major quality problems in engineering) is controlled separately or is sufficiently "mature" that all quality problems have been resolved.

Part 3 is applicable for supplied items of low intrinsic value where sufficient assurance of quality can be obtained by a rigorous final inspection which separates out "good" from "bad" products. In other words "stock" items of no great complexity.

Each part of the standard requires the essential elements of a quality system to be present:

A defined and documented system

Defined responsibilities

Formal reviews of the system

Documented work instructions

Adequate records

but the criteria themselves are given in such a general way that interpretation is not easy. For example, clause 4.6 "Purchasing" in part 1 covers the selection of subcontractors, purchasing data and verification of purchased product but all in a very general sense.

Also,

Clause 4.9.4 Inspection and Test Records in part 2 states:

> "The supplier shall establish and maintain records which give evidence that the product has passed inspection and/or test with defined acceptance criteria."

Both these clauses refer to very important aspects of a formal quality system—vendor assessment and records of the inspection of items. The criteria, as defined, are correct but they are obviously open to endless dispute about what *actual* policies and procedures meet the intent of these clauses. The answer is—it depends on the circumstances. A "general" standard has to be written in such a way that it can be applied to *all* types of supply of products and services so it can only state very general criteria. This will always be the case, whatever the standard, and it is no use expecting these quality system standards to be more detailed.

The real question at issue is—do these standards define *all* the necessary quality system criteria relevant to their scope? There are two major areas where, if care is not taken, they may not. If suppliers are working still to the 1979 version of BS 5750 part 2 (which many no doubt are) then the criteria do *not* define a total quality system for manufacture and installation as implied in the scope. In reality it defines criteria for an inspection system only. In other words, it is assumed that sufficient assurance of quality can be given by a close control of the inspection activities alone. This may be acceptable in many cases but is certainly not true for the manufacture and installation of significant and complex items in the offshore oil and gas industry. There are no clauses which cover, for example, special process control or work instructions—elements of a quality system which are essential to most manufacturing operations in this industry.

This inevitably means that part 1 or part 1 without the design control clause was really the appropriate standard in the majority of cases. This was a serious deficiency of this particular standard and you should therefore be aware that a quality system which meets the minimum requirements of the 1979 version BS 5750 part 2 *may* not be adequate. This deficiency is one which is recognised and is rectified in ISO 9002 (the comparable ISO standard) and therefore in the 1987 issue of BS 5750 part 2.

The second point is that, although the scope of the standard refers to the provision of services, the clauses in all three parts of the standard are relevant to the supply of products. It is difficult at first to see how BS 5750 as written (or indeed any of the other "general" quality system standards including of course the ISO standards) could be applied to services (eg inspection services, geological surveys, planned maintenance) which you *should* include within a formal quality system. This is a subject which will be discussed again in chapter 6.

THE NORWEGIAN STANDARDS NS 5801–3

The Norwegian national standards

NS 5801 Quality Assurance System

NS 5802 Inspection System

NS 5803 Basic Inspection

are stated to be based on BS 5750 part 1, 2 and 3 respectively and, in fact, the three levels of rigour of a quality system are approximately the same in the Norwegian system as in the British System.

In many ways, these Norwegian standards, although written as "general" documents are more easily interpreted and useable for the offshore industry than the British Standards. There are a number of reasons for this.

Firstly, the Norwegian philosophy defines "quality" as "conformance to specified requirements" so the whole emphasis in the standards is how to provide assurance that specified technical requirements have been complied with rather than to guarantee a rather nebulous level of "quality".

Secondly, the wording of the clauses is often more directly relevant to engineering design and manufacture than the wording of similar clauses in BS 5750 and is, therefore, clearer and more easily understood. For example, "inspection status":

clause 11 in NS 5801 states:

> "A system shall be established which ensures indication of a product's inspection status during all stages of manufacture. The contractor shall be able to distinguish between inspected and uninspected products by using some suitable means of identification such as stamps, tags, routing cards, move tickets or other check devices."

By comparison, clause 4.17 in the original version of BS 5750 part 1 stated:

> "The supplier shall establish and maintain a system for identifying the inspection status of material during all stages of manufacture. The supplier shall ensure the ability to distinguish between inspected and uninspected material by using some suitable form of identification."

It is fair to say, though, that the relevant clause 4.12 in ISO 9001 (and BS 5750 part 1 1987 version) does conform more closely to NS 5801 in this respect.

Thirdly, the three standards (NS 5801, 5802 and 5803) are written so that extracts can be made from one or more of the standards to suit particular purposes. The introductory section of each standard therefore states:

> "...However, this shall not prevent agreement being reached on parts from a standard or by selecting clauses from all three standards..."

This latter feature gives considerable flexibility in use and therefore it is appropriate to think of these Norwegian standards overall as providing the elements (or building blocks) for a quality system specification. This is a particularly useful aid to devising a quality system specification for a particular project or supplier *provided* you understand the criteria and can make a sensible selection of clauses.

In terms of the nature and content of the clauses of the standards they are, of course, similar to BS 5750 and other standards. The control and calibration of test equipment clauses are somewhat less specific than in BS 5750 and, perhaps, NS 5803 is more limited in its requirements than BS5750 part 3 but otherwise these Norwegian standards are broadly compatible with the equivalent British Standards.

THE ISO STANDARD 9000-9004

This standard has only (1987) just been made available and it will be a while before its application becomes general. This is intended to be the international set of standards for defining the criteria of a quality system and will, inevitably, possess the attributes of the various national standards in respect of the potential for different interpretations of the intent of the criteria. This, as was discussed earlier, must be accepted. If we look at the criteria themselves we see that there are four standards. ISO 9001, 9002 and 9003 define three different levels of quality system equivalent to, respectively, BS 5750 part 1, 2 and 3 and,in fact as stated, BS 5750 in the 1987 version is exactly the same as ISO 9001-3. ISO 9004, is rather different in that it defines quality management controls rather than quality system criteria.

The major difference in specific criteria (other than minor changes in format and terminology) between ISO 9001, 2 and the original BS 5750 parts 1 and 2 is that the ISO standards are more specific on the need for quality manuals and quality plans to be in existence. BS 5750 did not specifically state the need for them but it is unusual for them not to be produced in fact. ISO 9002 is more directed towards the idea of a quality system rather than, as BS 5750 part 2 was towards the idea of an inspection based system. ISO 9003 (the "final inspection" standard) is, likewise, a rather more quality oriented than inspection oriented standard but, in practice, this should not make too much difference for what is, in effect, a very simple quality system requirement.

The final standard in the series, ISO 9004, provides guidelines (rather than criteria) for quality management and quality system elements. It is a useful, although very general, document which provides possible interpretations of features such as documentation control, vendor assessment and so on. It also covers a much wider scope to include topics such as quality in marketing and cost considerations. It is not clear how the standard is to be used as it is hardly written in a format suitable for contractual purposes. It is more a general management guide/checklist of items to be considered when managing the quality operations. It is, therefore, a very valuable guidance document but it is not strictly a quality system standard in the manner of the other standards described in this chapter. Time will tell if this new international set of standards will obviate the need for the industry specific standards to any degree.

THE CANADIAN STANDARD CSA Z299

This standard is of great value to an industry such as this and it is a pity that it is not more widely used. Whilst being a "general" standard issued by the Canadian Standards Association and addressing the same quality system criteria as other standards, it derives directly from the quality assurance requirements of the power generation industry in that country. This means that the standard directly addresses the types of quality activity—planning, procurement, design assurance, manufacturing, inspection and test, etc—in the context of the complex, capital plant type industries. Therefore, the concepts and approach are particularly appropriate to the offshore oil and gas industry—at least in respect of design and manufacturing activities.

CSA Z299 has four levels. Level 1 is really a "nuclear" standard and as such is equivalent to the other nuclear quality system standards. Levels 2, 3 and 4 correspond approximately with BS 5750 parts 1, 2 and 3 (or NS 5801, 2 and 3 or ISO 9001, 2 and 3) respectively. The advantage of this standard is that it gives more detailed guidance on matters such as quality manuals and quality plans and the different types of procedure which need documenting—system function procedures (which are the management procedures), inspection procedures, manufacturing procedures and special process procedures. It also gives very explicit guidance on the categorisation of inspection equipment in terms of control and calibration requirements—in fact it defines four categories of instrumentation in terms of required accuracy, frequency of use and sensitivity to loss of calibration.

It also defines a method for logically establishing which level of the standard is appropriate in a particular case—ie a method of criticality analysis which is a valuable aid to the logical selection of the level of quality system needed.

C.S.A. Z299 Level 1.	C.S.A. Z299 Level 2.	C.S.A. Z299 Level 3.	C.S.A. Z299 Level 4.
B.S. 5882	B.S. 5750 Part 1.	B.S. 5750 Part 2.	B.S. 5750 Part 3.
ANSI N45. 2.			
	ISO 9001	ISO 9002	ISO 9003

← Increasing degree of rigour.

Fig. 2.10 Comparison of QA system standards

CRITICALITY ASSESSMENTS

Whatever the differences in the philosophy (or the aim) between the different quality system standards there are, in practice, only four levels of quality system which are specified (see Figure 2.10).

The degree of rigour (ie the level of assurance demanded) will naturally determine the cost of implementing the quality system. This we will discuss in chapter 7 but it is obvious that to implement a quality system which meets the requirements of a nuclear standard will need much more resources (and therefore will be more expensive) than to implement one to meet BS 5750 part 3 or NS 5803.

Therefore the selection of the appropriate level of quality system is an important decision which should be taken in a logical and rational manner.

Quality assurance programmes are applied because clients (and others) want assurance of quality and some level of confidence that products (or services) perform according to specification. One hundred per cent assurance is not possible (nobody is perfect) but the degree of assurance which is justified depends on two factors:

1. The probability of failure (this includes late delivery, product not performing to specification as well as component failure or breakdown).
2. The consequences of failure.

These two factors are not the same thing. For example, the probability of failure of a complex piece of machinery built to a novel design may well be quite high but the consequences of failure may be quite low because there is a duplicate unit which can be supplied in the event of failure. Conversely, the probability of failure of a lighting fixture could be quite low but, if it is part of an emergency lighting system on an offshore structure, the consequences of failure could be rather serious.

Therefore, it is helpful to try to make criticality assessments of components, structures, sub-assemblies, etc in order to devise the correct level of assurance needed and the correct level of quality assurance system to apply. Such criticality assessments can be relatively crude (there are only four levels of system to specify) and do not require detailed failure analyses (HAZOP analyses for instance). A very convenient basis for making these criticality assessments is the guidance given in the Canadian standard CSA Z299. A rating system (on the scale 0-4) applied to each of six characteristics of the project/component/item (the higher the rating the more critical the characteristic). These six characteristics are:

1. Design complexity
2. Design maturity
3. Manufacturing complexity
4. Special characteristics
5. Safety
6. Economics

The first three relate to the probability of failure and the last three relate to the consequences of failure. After grading each characteristic, the total score for all six characteristics is added up and, on the basis of this total score, the relevant quality system level is defined as follows:

18-24 means CSA Z299 level 1

13-17 means level 2

8-12 means level 3

4-7 means level 4

<4 means a formal quality system is not justifiable

The Nuclear Standard BS 5882 (the nuclear standard) also gives guidance on criticality assessments although not in a quantitative manner. Clause 1.1.4 of this standard lists some points which should be considered and these are:

The consequences of malfunction or failure

Design and fabrication complexity of novel features

The need for special control

The degree to which functional compliance can be demonstrated by inspection and test

The degree of standardisation

The difficulty of repair or accessibility for in-service inspection

These are really features which could help to define the rating of the CSA Z299 characteristics. In fact, it is not uncommon for clients or contractors to define their own criticality rating system for a given project or type of product.

This is perfectly satisfactory but, in truth, they are all more detailed explanations or elaborations of the six characteristics already mentioned. Their value is that they aid the selection of the correct rating for each of the six characteristics.

THE NUCLEAR QUALITY SYSTEM STANDARDS

These are all very similar (word for word the same in many areas) because they address a specific industry and its needs and can, therefore, be less ambiguous or open to interpretation than to general standards which we have just discussed. Since, as we have already stated, the nuclear power industry is similar to the offshore oil and gas industry in a number of respects—high technology, large and complex projects, safety critical items, similar design and manufacturing technologies—the nuclear quality system standards are worthy of study. In fact, if you removed the word "nuclear" from the standards, you would have a very suitable generic quality system standard for the offshore oil and gas industry—far more appropriate than the very general standards such as BS 5750.

The British Standard BS 5882 "Total Quality Assurance Programme for Nuclear Power Plant" is entirely typical. It contains many of the same elements as BS 5750 but is more explicit in some important areas. The "Quality Programme" and the "organisation" (clauses 1 and 2 of the standard respectively) are particularly explicit. The need for defining the programme for a project at the outset is stated together with the documentation and planning requirements. Likewise, the organisation needs to recognise the interdisciplinary nature of many quality activities and to recognise and control the interfaces between different groups.

An important principle which goes through the whole of this standard is the principle that inspection (or verification of quality in general) should be carried out by persons other than those who performed the actions which are to be checked. This is a very appropriate principle for nuclear safety related items as it provides the best and most sure (if not necessarily the cheapest) way of assuring the quality of an action. The same approach is appropriate for an offshore project because it shares many of the features of a nuclear project in respect of the seriousness of the consequences of a quality failure. This principle is stressed particularly in the clauses governing design control (the verification of design), documentation approval and inspection.

This nuclear standard also goes into considerable detail on the requirements for the control of procurement—particularly the relevant documentation and, indeed, lists ten criteria (or checklist items) for procurement documents. Likewise the identification of items is defined in detail with the emphasis (not seen in the general standards) on physical indications (tags, markers, etc) and the need for detailed traceability. This, again, is very relevant to the requirements of the offshore industry.

Finally, the stress on the controlling and retention of quality records is another area where detailed and explicit requirements are given which are not found in the general standards.

This discussion of the nuclear standards and the implication that they are more applicable to the offshore oil and gas industry than the "general" ones is not a way of saying that the *level* of quality assurance programme should necessarily be the same as in the nuclear industry. It is more a matter of approach. Remember, the general standards are written to cover *all* circumstances and are written to be used by *suppliers* of goods and services to demonstrate their commitment to the supply of quality products. The nuclear standards are written for application by *clients* who need to have assurance of quality of large, complex engineering projects. Inevitably, therefore, if you are an operator in the oil and gas industry then the style and format of, say, BS 5882 is more immediately relevant to your needs than the style and format of, say, BS 5750. This latter standard, as was stated earlier, needs considerable interpretation to be used. Hence, the view (perhaps heretical) that BS 5882, without the "nuclear" connotations, could be the model standard for *operators* to work to for the design and construction phase of a project. BS 5750 (and its equivalents) could be, when appropriately interpreted, the standard for the *suppliers* to work to in order to demonstrate conformance to the operator's (their customer's) requirements. This view leads naturally on to the question "How do you correctly interpret the requirements of a general standard?" Hence the "industry specific" standard.

INDUSTRY SPECIFIC STANDARDS

There are really two ways in which "industry specific" standards can be devised.

The first is to write "interpretational" or supplementary documents for use with the general purpose standards such as the ISO series, BS 5750 or NS 5801-3. This is the route taken by third party certification bodies (BSI and LRQA Ltd in the UK, for example) when they need to define the criteria for assessing suppliers in a particular industry sector (eg pumps and valves, compressors, castings, fasteners, structural steelwork). These documents are known by names such as "technical schedules" or "quality system standards" and they highlight specific test requirements or documentation control requirements relevant to the industry concerned. The necessity for these "interpretational" documents has been (or at least in the UK) partly a consequence of the deficiencies in the general purpose standards discussed earlier. The necessity for them in the future may well be not so great.

The second way is to write a specific quality system standard for an industry sector which is a "stand alone" document making no reference to any other quality system standard. The prime example of this in the oil and gas industry is the API standard Q1 "API specification for Quality Programs". As well as being the prime example it is, at the moment (1988), the only example specific to the oil and gas industry and this has only recently been introduced.

API-Q1 "SPECIFICATION FOR QUALITY PROGRAMS"

The aim of this standard is to provide guidelines for a supplier of hardware who wants to be licensed by the American Petroleum Institute to use the API monogram on his products. The criteria in the standard are therefore written specifically for the supplier of hardware and in such a way that he can be assessed as meeting the API requirements. Consequently, many clauses make direct reference to the necessity of meeting the technical requirements of other API standards.

For example, clause 3.11 "Traceability" states:

> "The traceability program, as required by both the manufacturer and the applicable API product specification, shall be specified in the quality manual."

This is a reasonable approach because the philosophy behind this standard is that it is to be used to assure the quality of design and manufacture such that an appropriate API technical specification is complied with. Clause 3.11 can be readily interpreted in conjunction with a technical standard because this will define the real tracebility requirements of material, parts and components. We stated (in chapter 1) that a product specification comprises both a technical specification and a quality system specification. The point about API-Q1 is that it is a quality system specification written to be directly compatible with particular technical specifications. It could not easily be used in conjunction with a non API technical specification.

The content of the standard covers, of course, all the major quality system elements (including design control). It would be fair to say that the level of system defined is closer to BS 5750 part 1 and NS 5801 rather than BS 5750 part 2 or NS 5802. There are a number of points of particular note, other than those already mentioned. Much stress is put on training and indoctrination of *all* personnel involved in quality matters (clause 3.3) and this is good. There is a requirement for a quality manual to be produced (and guidance on the contents is given in a non-mandatory appendix) and detailed emphasis is placed on design control, control of instruction and technical procedures, control of procurement of critical items and inspection and test, ie the areas of most concern in the supply of critical engineering components. There is less emphasis on "management" matters such as system review and the feedback aspects of a quality system. That is not to say that these features are not covered in API-Q1, merely they are not emphasised or covered in as much detail as the other features. API-Q1, to an extent, therefore emphasises the quality control aspects of quality systems. This is a characteristic of the special purpose standards (the relevant section of the ASME Boiler and Pressure Vessel code is quite similar in its philosophy). They are product and quality control oriented.

The general lack of industry specific standards is a consequence of the significant effort needed to produce them (and get industry wide agreement to them) so it is unlikely, therefore, that a sufficient number of such standards will ever be developed to cover all the needs of the industry. Likewise, it is unlikely that, at least in the short term, a sufficient number of interpretative documents (quality system supplements) will be devised to cover all the needs of the industry. Therefore there will still be a need for the general standards as statements of the fundamental criteria which will have to be subject to correct interpretation in each particular case. This puts considerable reliance on engineers/managers in all parts of the industry being aware of the "real" quality system needs and being able to translate them into practical systems which meet the intent of the criteria defined in the standards. Blind adherence to the letter of the standards is not sufficient; you must know what is really needed.

THE VALUE OF THE QUALITY SYSTEM STANDARDS FOR CONTROLLING OPERATIONS AND MAINTENANCE

All the discussion so far in this chapter has been about design, manufacture and installation. Clearly, the operational and maintenance activities should be as much subject to the control given by a formal quality system. In fact, BS 5750 covers the supply of products and services so it could be said to encompass these post installation activities. Furthermore, BS 5882 (the nuclear standard) specifically states that it covers "all phases of design, procurement, fabrication, construction, commissioning, operation, maintenance and decommissioning". However, to what extent are the particular requirements of these post-installation activities covered by the standards? It must be said, at the outset, that the standards have not, to date, been applied to any significant degree in this way. All we can say is that the same five major features will apply (as described at the beginning of this chapter) and we should expect that the quality system criteria in the established standards should also be applicable. Note the phrase "should be applicable". It may well be that for the quality assurance of operation and maintenance some criteria may be irrelevant and others may need amplification. What is necessary is that the requirements for the quality assurance of operation and maintenance activities must be logically analysed and the relevant clauses of the quality system standards matched against them. This we will attempt to do in Chapter 6.

DOCUMENTING THE QUALITY SYSTEM

In the early part of this chapter we discussed the need to develop a matrix of "features" and "activities" which effectively defines the quality system. Later on, in discussing the various standards available, mention was made of the requirement to define and document policies and procedures. Clearly, the documenting of a quality system correctly is a key first step in implementing one.

In each of the next three chapters which cover specific major activities (design and procurement, manufacture and installation, and certification) the specific quality documentation needs will be addressed but there are some general points worth making at this stage.

The documentation of a quality system serves two purposes:

1. It produces authoritative guidance for staff so that they perform their activities in a consistent and correct manner.
2. It provides demonstrable evidence that a quality system is in operation.

A quality system, be it for the overall operations of an organisation or for a specific project or contract, requires three levels of documentation as shown in Figure 2.11.

At the top level we have the quality manual which should be a relatively brief document which defines the organisation, the responsibilities in respect of "quality" and the quality policy to such a degree that it is clear that the criteria set (be they in a quality system standard or in a client's specification) are covered. This will require reference to the existence of specific management procedures. A good, general guide to the preparation of a quality manual is found in references (2.1) and (2.2).

At the next level down there should be a number of management procedures which define in some detail the carrying out of the various activities (eg vendor assessment, non-conformance reporting and control, design change control). The procedures should be written in sufficient detail to ensure that they are working documents. The acid test is: could another member of staff (of the appropriate level) carry out the task correctly by working to the written procedure?

At the final level there are specific technical procedures (eg inspection

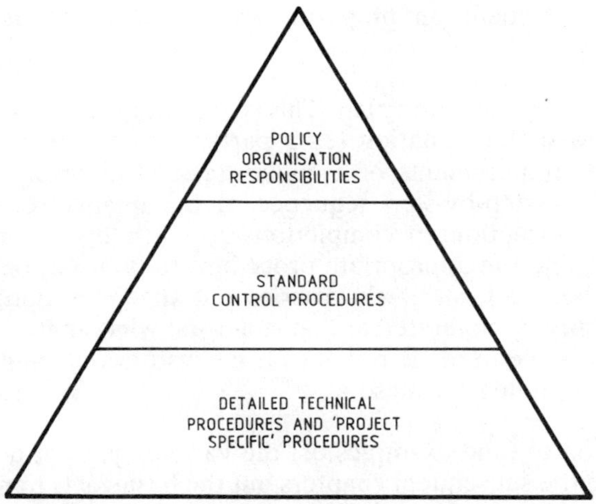

Fig. 2.11 The three levels of quality documentation

procedures, welding procedures, equipment calibration procedures) which must define, again in sufficient detail, the activities so that a competent (ie appropriately trained and experienced) person can carry out the tasks based on the procedure.

The provision of all this documentation is a significant task but it is, in itself, a good discipline as the writing down of a policy or a procedure requires you to define very carefully and unambiguously what you require to be done.

Apart from these three basic types of documentation there is often, particularly in "project type" design and manufacture a need for a further document called a quality plan. The terminology is often imprecise and there is some confusion between a true quality plan and an inspection plan.

A quality plan is a document, specific to a project (or part of a project) which defines how the particular quality requirements of the project are to be achieved. In contrast with a quality manual (which defines overall quality policies) and procedures (which define specific operations), the quality plan is the matrix or compilation of procedures and policies which show exactly how a project's quality requirements are to be met. In other words there should be a cross-reference from each item in the client's (or contractor's) specification to a set of procedures and a section of the quality manual (chapter 4 gives an example of such a document).The plan should identify any "hold" points and the input of other parties (eg clients or inspectorates). The idea is to provide a master document, making reference to the existing quality policies and procedures, to guide the quality activities of the project. This can be seen, therefore, as the "core" planning document in a quality sense and effort spent on preparing such a document is always well worthwhile.

Allied to this is the inspection plan. This is sometimes also called the quality plan and, for a simple situation (eg a particular item or product), it can encompass the requirements of a quality plan. This inspection plan is a chronological or step-by-step sequence of the quality related activities required from inception to completion. At each operational stage the plan should define the appropriate procedure to be used, other reference documents where relevant, who carries out the operation, what documentation (if any) is generated at that stage and what analysis or review of the operation is required. It provides a means of controlling the quality activity in a chronological sense.

More discussion of (and examples of) the various types of documentation is presented in the subsequent chapters but the basic facts to remember are the two prime purposes of quality system documentation and the restraint needed to prevent over-documentation of the system.

REFERENCES

2.1 *Guide to the preparation of a quality manual.* The Institute of Quality Assurance. London 1979.
2.2 Nicholson, S. Problems associated with the production and implementation of quality plans. *Journal of the British Nuclear Energy Society*, July 1977.

CHAPTER 3

QUALITY ASSURANCE IN PROJECT MANAGEMENT, DESIGN AND PROCUREMENT

J. G. Hicks

INTRODUCTION

Project management, design and procurement are activities which it is appropriate to group together as they are the engineering, planning and control aspects of a project, as distinct from the manufacturing and installation aspects. Although it is, perhaps, something of an over simplification, they can be thought of as software, rather than hardware type activities.

They share two further characteristics which are important in the context of quality assurance. Firstly, they are activities which are not immediately seen as amenable to the rather manufacturing based concepts of the formal quality system approach (as typified in the standard criteria discussed in the previous chapter) and, secondly, they are activities where failure in quality can have a large "knock-on" effect. As an example from a different, though in many ways, similar industry, the Central Electricity Generating Board estimated (ref 3.1) that the ratio of design effort (ie extra work to ensure the quality was adequate) to savings in site effort (resulting from correct and appropriate design) was of the order of 1 to 100. In other words, an extra £10,000 spent at the design and planning stage can save one million pounds on site costs.

We can, therefore, say that the aim of quality assurance in project management, design and procurement is to achieve project completion to the correct quality as efficiently as possible. This, you might consider, is a fairly obvious aim but if we analyse it we can see that its fulfilment requires two types of "control".

First of all, the engineering or technical requirements of the project must be correctly defined and implemented. This is fundamental to the whole issue in translating an operational requirement (however well or ill-defined) into a set of correct, complete manufacturing instructions which, when followed, produce the required installed hardware.

Somewhat separate to this are the management requirements of the project, which must also be correctly defined and implemented. These, in so far as they relate to the quality management of the project, effectively set out to ensure that the engineering or technical goals of the project are achieved and can be demonstrated to have been achieved.

It is this second type of "control" which we are concerned with in this chapter. The various management (as opposed to strictly technical) activities of project management, design and procurement in the offshore industry will be analysed in terms of their quality assurance implications and how they need to be organised (and documented) to demonstrate that formal quality assurance is being applied. This implies, of course, that the formal quality system criteria defined in the various standards (see previous chapter) can be readily interpreted in this area of operation. This is not immediately clear, so the first step is to look at the conventional quality system criteria and see how relevant they are likely to be to project management, design and procurement.

THE IMPORTANCE OF THE CONTRACTUAL RELATIONSHIPS

Throughout this chapter we will be concentrating on and discussing what are, essentially, the contractor's actions (rather then the manufacturer's or the operator's) as it is he who normally takes the initiative in devising and implementing quality systems for a project. It is usually the contractor, after all, who has the prime responsibility for assuring the quality of the project and demonstrating to the operator (the ultimate client) that his requirements have been complied with. Of course, in a large project, the operator may well run a formal quality assurance activity himself to match the contractor's operations. Likewise, subcontractors and manufacturers will operate formal quality assurance to match the contractor's efforts. The contractual relationships and the responsibilities of the different parties to a project are complex and likely to differ from project to project, although table 3.1 indicates the likely relationships and responsibilities, in respect of "quality" matters, that are likely to be found, certainly in a large project.

This complex contractual problem inevitably leads to a need for great attention to be paid to communications and coherency of approach amongst the parties concerned. The overall planning and the interfaces between the different quality plans or quality programmes of each party in the project are therefore crucial. In this context, it is important to recognise the slightly different perceptions which different parties will naturally have about the type of (and degree of detail of) quality assurance measures. Everybody

TABLE 3.1 Typical Relationships and Responsibilities (Simplified)

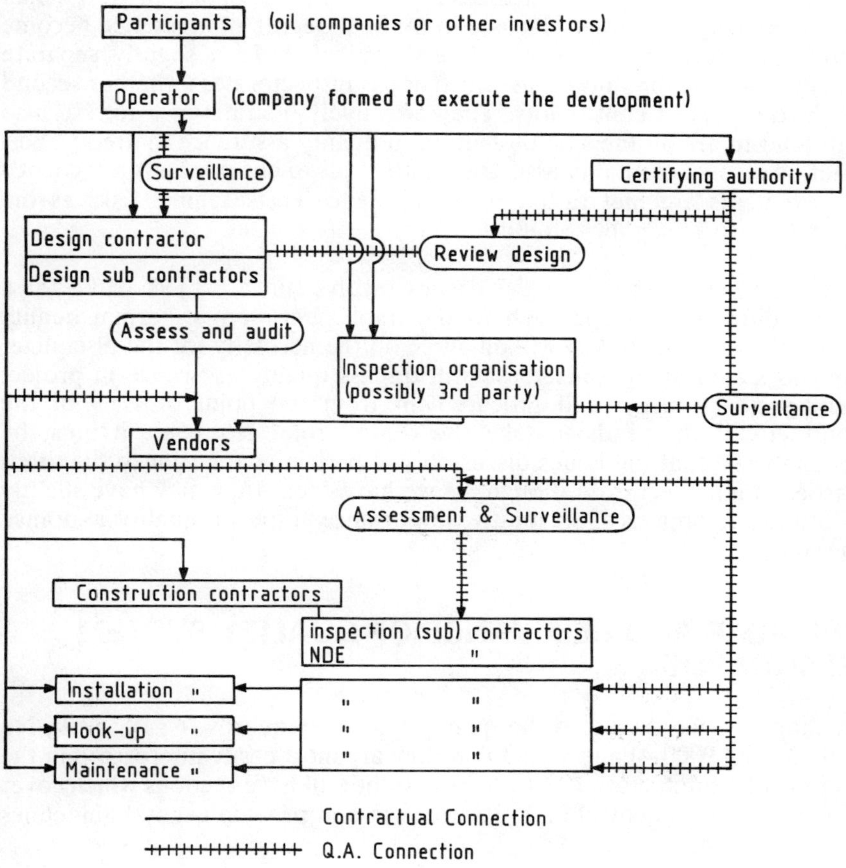

wants a "quality job" but inevitably the more detailed and extensive the degree of assurance which is required (or imposed) the more the quality assurance activities will cost. (The final chapter gives some detail on the magnitude of such costs.)

The balance of costs and benefits of a quality assurance programme may not be seen in exactly the same way by all as the contractual responsibilities and financial exposure will not be the same. For instance, the operator is, in the end, totally responsible for the installation and will ultimately bear the costs (particularly the potentially large consequential costs) of any failure in quality. He will tend to err on the side of requiring more rigorous quality assurance measures as the cost of them is very small in relation to the risk he is running. At the other end of the spectrum, the suppliers (particularly the smaller suppliers some way down the contractual chain) are concerned about the cost of quality assurance, as it can be a significant part of their overall costs. Inevitably, they cannot afford to err on the side of excessive quality assurance measures as, from their point of view, small and, perhaps, negligible increases in the real degree of assurance become proportionately more costly and less justifiable. In a slightly separate situation are the organisations which act as inspectorates (whether second or third party) or consultants. They effectively operate on a fee basis so will tend to err on the side of caution in quality assurance matters. Their brief, in respect of quality assurance matters, is to ensure that their client's requirements are met rather than to balance costs against risks as the operator or the supplier would.

These differences between the parties involved in a project make for a subtle difference in approach to the implementation of formal quality policies even though they will all agree on the necessity for the discipline. As was stated at the outset, we will discuss quality assurance in project management, design and procurement from the point of view of the contractor as he tends to take the central role. However, it must be remembered that the issues discussed will be just as relevant to the other parties in a project even though, as we have seen, they may have slightly different perceptions of the relative importance of specific quality assurance measures.

THE ROLE AND RELEVANCE OF QUALITY SYSTEM STANDARDS

In chapter 2 we discussed the quality system standards (eg BS 5750, NS 5801-3, ISO 9000-4) and stated that they are most easily interpreted in the context of manufacturing. Nevertheless, they all have sections which cover planning, design control and procurement and provide general guidelines for these areas.

Planning

Planning, as defined by the quality system standards, relates to the planning of quality activities. The major activities covered are the production of quality plans, the proper defintion of quality requirements and the assurance that the appropriate equipment, facilities and skills are available.

These are necessary activities but are not specifically related to project planning as the latter topic is of a much more comprehensive nature.

Design Control

This topic is covered in some detail in the quality system standards and, as was discussed, in chapter 2. All the high level Standards (eg BS 5750 part 1, NS 5801, CSA Z299 level 1) include a clause which covers design control but, of course, other clauses in the standards (those relating to records, corrective action, documentation and change control, for example) are also relevant.

The phraseology of the different standards, naturally, varies but all require certain major management steps to be taken to assure the quality of a design:

Firstly, design should be carried out according to established company procedures and be properly documented—ie it implies a company code of practice for design.

Secondly, design work should be carried out by competent, qualified personnel and the design outputs verified by personnel other than those who did the original design.

Thirdly, any design approval should be carried out by specific procedures (independent analysis or test work, for instance) and properly documented.

Fourthly, design reviews should be undertaken to evaluate the suitability of the design. These reviews should be formal exercises carried out by non-design people.

Fifthly, all design changes should be controlled in the same way as the original design work.

Procurement

Again, all the major standards say very similar things even though the phraseology differs. The major management steps are:

Firstly, the control over the definition of what requires to be purchased—ie getting the purchase documentation correct and authorised.

Secondly, the selection, approval and control of suppliers—eg maintaining a list of approved suppliers or operating a vendor audit programme.

Thirdly, the checking of the quality of incoming supplies (either by physical inspection, review of supplier's documentation or reliance on the supplier's quality system).

Fourthly, the identification (and segregation, where necessary) of incoming goods as being satisfactory, defective, suspect or awaiting inspection.

These are fairly straightforward concepts and are easily applied in the general manufacturing area but are not necessarily easily applied in the procurement phase of a large project.

In general, we can say that the quality system standards provide a framework for the quality assurance of project management, design and procurement but they cannot readily be interpreted to produce specific guidelines for a quality system. In other words, the clauses of the standards are sensible and correct but do not, in themselves, give practical guidance for the user. What we must do, therefore, is to consider, in turn, each of the management activities in project management, design, procurement and see what is required to provide the right degree of quality assurance. As stated earlier, we will look at this primarily from the point of view of the contractor but bearing in mind the relevance of the other parties likely to be concerned.

In discussing these management activities, it should become clear that they do, in fact, fit into the overall requirements of a formal quality system as defined in one of the high level standards (eg BS 5750 part 1 or NS 5801).

PROJECT MANAGEMENT AND DESIGN

The Role of the Project Manager

Project management encompasses the activities of planning, resourcing, direction of effort, and control of progress and cost. The project manager is the driver of the project and in some companies is responsible for preparing the bid which, if successful, will become the project. In this way, if the bid is successful the project manager gets off to a flying start; knowing much of the basis of the task he is able to set the project in motion with a minimum of delay and misdirected effort.

The project manager may not be involved with the detailed engineering decisions where there is also a project engineer in the team but his role in controlling the work is vital to the accomplishment of the contracted task, be it design or construction. Project management is essential for the project engineer to pursue his job effectively and to the programme,and so project management must be considered a quality related activity. That is the argument for bringing it within the scope of a quality programme.

Contract Review

The preparation of a bid will have included at least an outline of the project plan indicating the disciplines and their activities on a timebase. This much is necessary for arriving at a price. On commencement of the project, a more detailed plan can be constructed incorporating subsidiary plans for individual activitities.

There is a risk that the scope of the work, on award of the contract, may be different from, or more detailed than, that used in arriving at the bid price. The rush in which most bids are prepared also leaves the possibility of some item being left out by mistake. A device known as a contract review is, therefore, employed before work gets seriously under way. In this review, all the senior personnel examine the contract documents in detail to ensure that the team has the right resources for the work. This is directed, particularly, at the disciplines needed, special facilities such as computer software, specialist staff or consultants needed and any unusual statutory requirements.

The contract review must be thorough and properly reported so that any actions found to be necessary are picked up by those in a position to effect them. The quality assurance activity relating to this comprises, firstly, ensuring that a review is built into the programme and that a standard procedure exists in the company's Quality Manual and, secondly, ensuring that the review is conducted and reported on by the personnel nominated in the procedure in the manner described in the procedure.

This, clearly, is part of what the quality system standards refer to as "planning". It is important that this planning stage is properly reported and documented.

Organisation

By organisation is meant the means by which personnel are put into a hierarchical structure through which authority can be exercised and response conveyed. The structure also identifies routes across boundaries of disciplinary groups or departments. Each structure must be designed to fit the project in hand but there are some points of general application which should appear in any organisation. The first and, perhaps, most important, is that there can be only one person responsible for running the project whether he be called project manager, project engineer or project leader (in the rest of this chapter he will be called the project manager). The second point is that there should be only a limited number of personnel reporting to the project manager. The actual number will depend on the nature and scale of the project and there are various theories of management which offer guidance. Typically, five is convenient for day to day operations but in a large, interdisclipinary team it may be necessary to increase this.

Quality assurance involves seeing that the organisation, as represented on a chart for the project, is how the personnel really are organised. It does not involve seeing that all the right disciplines are there. That is the project manager's job following on the contract review. The organisation chart, called an organogram by some, must show the total hierarchy in terms of the position, title and clear lines of responsibility. A particularly important point is that each position box should contain the name of the person fulfilling that role. A. N. Other is not good enough; if the project requires people in the organisation, they must appear. Without real people, the organisation does not exist and the work cannot be done.

An example of an organogram is shown in Figure 3.1.

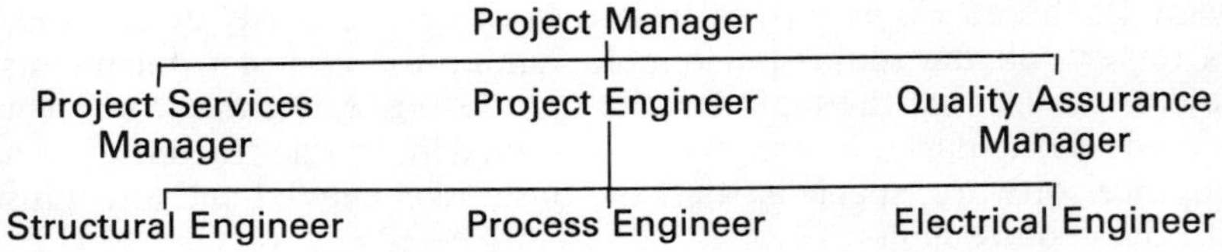

Fig. 3.1 Simple organisation chart or organogram

Job Descriptions

The job descriptions are complementary to the organisation chart. They describe the duties of the incumbents of the named positions and also the person to whom they report and the persons for whose work they are responsible. Writing job descriptions is not an easy task. A comprehensive description must give the scope of the work in all its aspects but must avoid incorporating matters which are properly conveyed in the operating procedures. A sound job description, therefore, describes the professional and administrative duties of the job holder without describing the mechanics of execution. An example of a typical job description is given here in table 3.2.

The organisation and the associated job descriptions are, as is obvious, necessary in any event but in so far as they relate to the "quality" activities (which they will always do to an extent) they must be defined and documented to demonstrate conformance to the "organisation" clause of a quality system standard.

Project Administration

This phrase has a restricted meaning in the sense that it encompasses the activities by which the project management is executed. These deal with matters of correspondence, filing, time and cost records, insurance, personnel, office accommodation for the project team, living accommodation, transport and all other matters relating to the management of the team. If the project is being conducted in the permanent offices of the company, many of these tasks will be performed by the company administration. In a site situation the work may be performed by a Project Services Group under a project services manager who will be part of the project team.

It will be understood that many of these functions are not usually seen as being "quality related" and will not necessarily appear in the quality plan for the project. Nevertheless, there are some functions which must receive the attention of the Quality Assurance Department and, of these, the most likely candidate is correspondence control. It might be considered that this is part of the document control activity but the nature and the variety of

TABLE 3.2 Example of How a Job Description Can Be Documented

JOB DESCRIPTION

POSITION:	Lead engineer—topside structures
REPORTS TO:	Project structural engineer
RESPONSIBLE FOR:	Senior designers, structures

The Lead Engineer is to:

develop from the conceptual design the detail design of all the topside modules

identify the structural reliability criteria from the project specification and select the software to be used

liaise with the Certification Engineer in respect of document packages for submission to the Certifying Authority

act on information received from the weight control department in order to achieve total budget weight and control CG migration

effect the distribution of effort within his group and monitor the performance of the senior engineers and their intermediate and junior staff

maintain daily contact with the Project Structural Engineer and report on trends relating to manpower and cost

nominate engineers to check calculations

allocate responsibility for the drawing register

sources of external correspondence merits separate attention. The significant features of a correspondence control system are that incoming and outgoing correspondence should be recorded in a register and that incoming correspondence should be directed to the person who is concerned with its content. The system should include a facility whereby no incoming letter remains unanswered for more than a certain time without a signal being sent to the recipient of the letter. A similar, but more urgent, reminder system is required for telexes and telefacsimiles.

More detailed consideration of some types of correspondence are given under the next section.

Project Communications

Sound project execution requires that not only are relationships between personnel and parties defined, as in the organisation chart, but that the mechanisms for effecting communication are available. Communications in this context are between the members of the project team as well as between the parties to the contract, ie the client, design contractors, construction contractors, vendors, certifying authority and any other party to the contract.

Some types of communication involve transmitting information on standard forms such as drawings, calculation sheets and requisitions. The conventional way of doing this is to prepare a covering sheet for the batch of documents called a transmittal. This has a standard distribution list printed on it and spaces for the title of the documents. The transmittal is sent with the documents in several copies, of which the document controller may keep one and the recipient keeps another.

This system is frequently used within the project team and may also be used between the parties. In the latter case, although the format may be similar, it is not wise to send transmittals to individuals. They should be addressed so that they will be processed by the document controller of each party. A system which degenerates into cosy little notes between individuals is a signal to the quality assurance surveyor that scope for failure in communications is developing. History has far more examples of engineering failures through poor communications than through any technical cause.

Letters have fallen into virtual disuse as a means of project communications because of the speed and availability of telex, computer communications and facsimile. The speed with which these can be sent implies a swift answer but this must not be to the detriment of a sufficient circulation. This must include all those in the project team who need to know the contents and respond to them.

Despite the advances in telecommunications, personal contact remains a prime method of communication. Circumstances surrounding a particular matter are not readily appreciated from a telex, and personal visits can be very important. A regular visiting schedule can be helpful as long as the frequency is adjusted to be compatible with the demands at any time. The personnel making these visits should be from the appropriate level in the organisation to the activity under review. It is important that minutes are kept of the substantial matters discussed, observations made and actions required. These minutes must be distributed to those who may need to be aware of the situation and suitable arrangements made for monitoring their responses.

Document Control

The function of the document control centre in a design organisation is to take documents from the originator and issue them to the nominated recipients; amendments are treated likewise. The intent being that at all times correct, complete and up to date issues of project documents are held by the appropriate persons. This is the simply stated aim of "documentation control" but, particularly in a large project, achieving this requires carefully designed procedures and painstaking attention to detail and monitoring of the system. Besides this distribution function, the document control centre maintains registers of the documents recording their status and is therefore in control of the issue of numbers. In a large project, the function is computerised and can be a complex set of activities working across boundaries between contracting parties as well as being an internal organisation. Absence of a sound document control system can lead to engineering errors arising from lack of up to date information.

The interest of the quality assurance department in all these communications is served by seeing that procedures for communications exist and are implemented.

Project Execution

Progress is not achieved from the outset by the project manager gathering the team together on day 1 and saying "Let's go to it!" The way in which the project is to be pursued must be carefully considered and all the choices examined and assessed. Having done this, the Project Manager can then write the project execution plan which, in effect, tells the team what their job is and how to go about it. It conveys to all the senior project staff the basis of the design and construction and any unusual characteristics of a chosen solution. From the project execution plan will emerge the networks for all activities with development into more and more detail. This will include the quality assurance activities as well as the quality related activities. At this point, the activities which are the specific responsibility of the quality assurance group must be identified. It is very easy for the picture to become muddy as different people pick up different ideas about what the quality assurance role is. It is then a sound idea to hold an educational session very early in the project's life in which the project quality assurance engineer explains the role of quality assurance. In this, he should drum into the project team that they are the people who are responsible for the quality of the work and that he is responsible for seeing that the proper quality management systems are installed and working.

A simplified example of a project network is given in Figure 3.2 showing those activities which are, typically, the responsibility of the quality assurance group.

Standard Procedures

The purposes of standard procedures are:

- to record methods of working which have been found to be effective
- to act as instructions to personnel new to the work
- to form an instrument which conveys to personnel working in adjacent activities the possibilities of interfaces with their own activities
- to act as a basis for review and improvement of execution of the operation

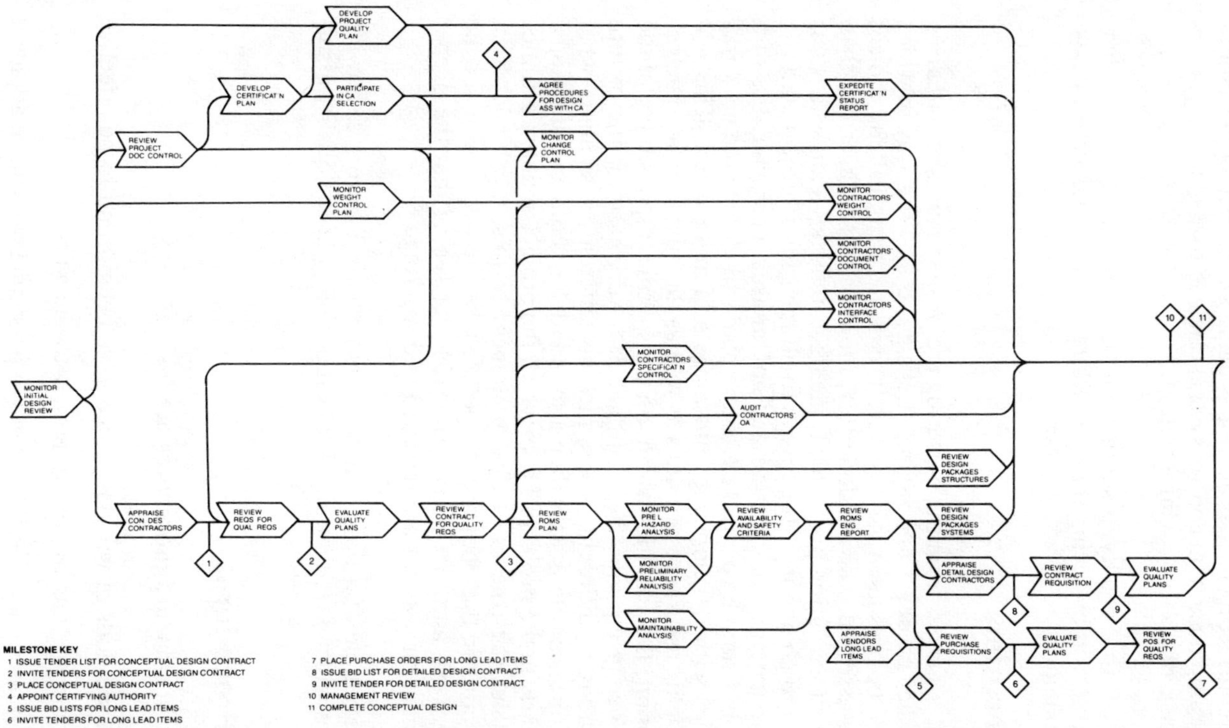

Fig. 3.2 Quality activities conceptual design phase (from reference 3.2)

In the case of new procedures:

> to act as a basis for development by requiring that the operation be thought out and written down for subsequent review

For the quality assurance surveyor, a standard procedure can be the instrument against which an audit is conducted. It is also a specific requirement of *all* formal quality systems that "procedures" be defined and documented to an appropriate degree.

The importance of the standard procedures to the quality system of a company cannot be over emphasised. They represent the *only* means by which an auditor can make an objective judgement of the conduct of the work. It is, therefore, worth spending some time considering what standard procedures should contain and how they should be compiled and administered.

Referring back to the purposes listed at the beginning of this section, it is important that these documents say how things are done, not how the company would like them done or how the company thinks the client would like to see them done. One of the difficult things about writing a procedure is to decide what is properly a procedure activity and what is within the normal skills to be expected of a person executing that procedure. People differ in the way they approach the details of the task and the procedure must be sufficiently flexible in allowing for this without compromising the aims of the activity. Recognising the above points, it has to be concluded that the only person who can write a procedure is the person who implements it. When this practice is followed, the procedure actually represents what happens. If this is not what is considered correct by the supervisor of the activity, then one value of a written procedure becomes apparent, namely that having it written down makes errors detectable.

Standard procedures can be at several levels of use. There are those which are used at all times for all tasks, others are used only for certain types of job whilst a more restricted group may be written for one particular job. The latter are required by a number of offshore operators because they regard most company house procedures as inadequate, perhaps with some justification, but not without some risk to the quality of the work. By definition, a newly written procedure does not represent a tried and tested method of working and there is no record of quality arising from that procedure. Another hazard is, that it has been known to take so long to write and obtain approval for a procedure, that the job has been finished before the procedure has been issued! The relationship between these documents is shown in Figure 3.3.

There is a need for direction in the performance of some activities which do not lend themselves to rigidly enforced procedures, for example, laying out of details on a drawing sheet. A type of presentation which can be used

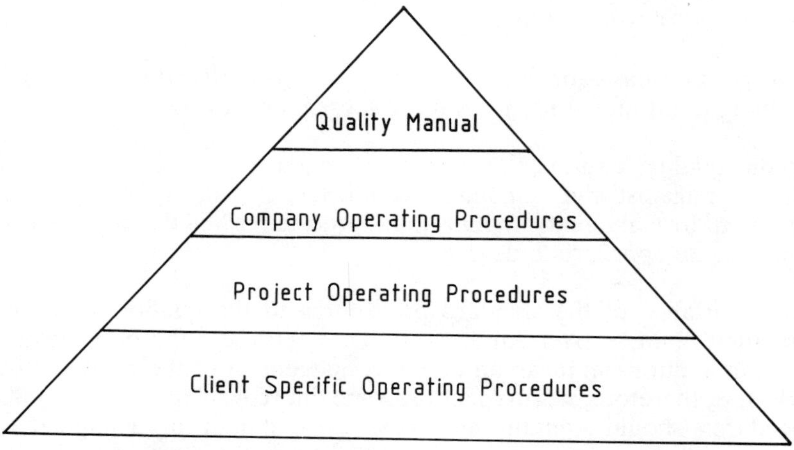

Fig. 3.3 Relationships between procedure documentation

here is a standard guide. This is prepared, controlled and amended in the same way as the standard procedures but failure to work to it does not constitute a non-conformance. A third type of standard is the specification. In the oil and gas industry, there are client, industry, national and international standard specifications. Nevertheless, some design contractors find it necessary to prepare their own specifications for areas not served by others or for clients outside the mainstream of the conventional industry. It is important that these are maintained at a sound technical level and do not conflict with other standard specifications on the same subject. They must, therefore, be subject to a regular review. It is the practice in some companies to have a standards review board which meets at planned intervals. It examines the company's standards of all types and approves drafts of new standards which it considers suitable.

Standard procedures should ideally be of a common format (including the control pages). The contents should be put under a sequence of standard headings of which table 3.3 gives a typical example.

The procedure document should commence with a list of the contents 1 to 7 etc. It is helpful if the text is generously spaced, especially between sections; to do so makes it easy to incorporate amendments to the text without changing the format extensively and facilitates the substitution of amendments as single pages.

Quality Plan

The quality plan for a project is really the "core" document from a quality assurance point of view and its preparation, review (and amendment where

TABLE 3.3 Suggested Layout of a Standard Procedure

1. SCOPE

This defines the overall operations to which the procedure applies; eg the design of carbon steel pipework for offshore installations.

2. APPLICATION

This points to the specific activity which is the subject of the procedure and any limitations or exclusions; eg this procedure applies to the use of CAD for the design of offshore pipework in non-hazardous areas.

3. RESPONSIBILITIES

This defines the person responsible for the execution of the procedure; eg the lead piping draughtsman.

4. RELATED PROCEDURES

Other company standard procedures impinging on or referred to in this procedure are listed.

5. REFERENCE DOCUMENTS

Documents referred to in this procedure other than related procedures are listed; eg national standards, data bases, proprietary software operating instructions.

All the above items can be seen as the information needed to establish that before the task is begun:

—the relevant standard procedure has been selected

—the correct person is executing the procedure

—the appropriate background information is to hand

Next comes a section which defines terms used in the procedure:

6. DEFINITIONS

Many can be included with the minimum of effort by referring to documents such as BS 4778 for quality matters. Other terms specific to the procedure will have to be defined clearly but concisely or possibly by reference to other documents.

Finally, the actual operation which is the subject of the procedure is described; eg:

7. USE OF THE CAD SYSTEM FOR DESIGNING PIPEWORK

---

necessary) are fundamental actions in the quality assurance of a project, as we saw in chapter 2. It is, in some ways, a subset of the main project plan as it defines, in some detail, the quality activities, who is responsible for them and how they are recorded.

Depending on the size of the task, the quality plan may be a simple bar chart or a complex document running to many pages with schedules of activities, networks etc. It should be appreciated that even a single platform installation represents an enormous investment and activities by numerous companies which may be conducted on a world wide scale. The major oil companies have found that they can prevent some of the large quality losses of the past by strict attention to quality management in design as well as construction.

The quality plan lays down the actions which are taken in the pursuit of quality in relation to the execution of the project. They will appear on the project network on which some of them will represent hold points beyond which work cannot proceed, without approval by the quality assurance manager for the project.

The quality assurance tasks appearing in the quality plan may include:

preparation of the project quality plan

preparation of quality assurance procedures

software quality audit

appraisal of design sub-contractors

quality system audits

appraisal of vendor quality plans

Project Progress and Reporting

Management of any activity cannot be effective unless the manager knows whether the activities are on time and budget according to the established plan. Accordingly, there must be an arrangement whereby the state of completion of any activity is reported and recorded in a form which can be rapidly assessed by the project manager. This can be done by the basic device of the achievement of each identified activity being reported at regular intervals. This is not a complicated matter if it is a measurable achievement, say in terms of the number of drawings or requisitions issued or tons of steelwork delivered. It can be more difficult in a less well quantified activity, such as process design or structural optimisation. In that type of activity, the responsible engineer must arrive at an estimate of the percentage accomplishment of the task. Since this reporting is progressive, each report reflects the up to date position and absolute correctness at any one report date is not essential. It will be apparent that the accuracy of this type of report should be better towards the end of the task than at the beginning. Any large swings towards, or away from, completion at the later stages indicate that something is out of control, and attention by the project manager through a design review may be necessary.

Most projects of any size will employ a computerised system of project control into which the reported state of any activity will be entered. The output will be in some form which will enable the project manager to assess where attention needs to be paid to correct any slippages or overspends.

The quality assurance staff must play their part in establishing the project plan and in its execution for although most of these matters are not the concern of the quality assurance group, they are of relevance inasmuch as they reflect their own activities. A slippage in the audit programme, for example, can be seen as a quality system failure and therefore, warrants attention.

Design Review

The design task is pursued along the lines laid down in the project execution plan and various activities such as interdisciplinary checking and interface control are exercised. These activities are somewhat executive in their conduct and are not planned to examine the route being followed. Experience has shown that there is a need for external examinations of design activities for a number of reasons. At the most basic level, these examinations, or design reviews, can ask the question—is this the best solution for this circumstance? Are there others which have not been thought about or some which have been discarded? This may confirm that the chosen solution is indeed a sound one or it may redirect efforts.

At the next level of review, it may be found that, although the various solutions being pursued by different disciplines were sound individually, they were not good as a total solution. The disinterested body of engineers may then generate alternative approaches in some areas.

Perhaps the ultimate level of design review occurs when it is found that the original objectives no longer exist in the form for which solutions were designed. A simple example of such a circumstance was the sharp drop in oil prices in 1986. Designs for marginal field installations became completely unprofitable almost overnight. This change was so severe and so rapid that it didn't need design reviews to show that the contemporary solutions were no longer viable!

It is necessary that design reviews are carried out in a systematic manner and are properly documented and the agreed follow-up action maintained and recorded. This implies a design review procedure, properly minuted meetings and records of actions taken. Design reviews are undertaken to prevent, as far as possible, expensive design changes being made late on in the project. The cost of a design change probably increases exponentially with the stage of the project at which it is made. Consequently, the design reviews early on in a project before the design is "frozen" are of great importance and should not be carried out in a superficial or informal way. The contract review can even be thought of as the very first design review stage.

The design review should, therefore, be conducted by a review board consisting of the design contractor's Chief Designer with the assistance of heads of discipline departments and in-house and external specialists. The project manager, project engineer and other project staff relevant to the area under review are also present. Depending on the nature of the subject, the quality assurance manager may attend.

The approach taken in such a review should be constructive and the project manager can use the opportunity to bring the assembled knowledge and experience to bear on the job. From the quality assurance aspect, the points to observe are that the review was as planned in the project plan and that it was conducted and recorded in the manner set down in the approved procedure. Naturally, if any quality system non-conformances are exposed, these should be examined and corrective action implemented.

Technical Audit

These are sometimes confused in name, at least, with design reviews. They should be much more confined to examining the techniques and data being employed in the work. Such an audit will examine one part of the design and seek to assess the validity of the approach. It is not a quality assurance function but its planning, execution and reporting should be confirmed by quality assurance observations. If any quality system non-conformances are revealed, then appropriate action should be to correct them.

Quality Assurance Administration

In a project, the quality assurance personnel are subject to the same organisational disciplines as the other members of the project team; their position and responsibilities are defined on the project quality plan. There must be a clear distinction between the quality assurance, quality control and the quality related functions.

Quality assurance functions must appear on the project network with the timings related to all the other activities. They must be costed on the basis of the man hours which should, as far as is possible, be estimated on planned activities. This type of management information is often set out in the form of a cost, time and resource summary, CTR for short. This is given an identification code so that it fits in with all the other CTRs in the project and forms part of the total project management information.

Close Out Reporting

In a design contract, the output, or "deliverable" as the jargon has it in some quarters, consists of a description of the artifact designed. This description is, primarily, in the form of engineering drawings but usually requires a great deal of written matter also. The client will usually require a report on the basis of the design and its execution, and any conclusions about the particular features of the design and its execution in relation to the original requirements. Its preparation is an essential discipline for the design team and a means whereby the client can have the validity of the solution checked by his own staff or other body.

The role of the quality assurance department in this area is to ensure that the material is produced and reviewed by the personnel designated for this purpose in the operating procedures.

SPECIFIC DESIGN CONTROL ACTIVITIES

Project management and design are, to some extent, interwoven in a quality management sense but there are some matters relating specifically to the design task which must also be considered.

Preparation for Design Work

Before a start can be made on any design activity, there are some basic sources of information and systems to be appealed to or established. One of the most fundamental is the design database represented by the job specification, standard codes of practice and standard specifications. At the most elementary level the availability of this information as hard copy must be established. At an engineering level, the software to be used should be verified as including the appropriate data and any new data loaded in the software.

The outputs (deliverables) will be in the form of drawings and written documents and it is important that the format of these should be established from the outset. In addition, units and scales should be defined before any work is attempted; the scope for errors in this area is truly enormous and it is, therefore, a matter of extreme importance. Associated with this function is the need to prepare guidelines for the assembly of engineering specifications and data sheets, which may, eventually, be used as part of the procurement package. At the same time, there is a need for an equipment numbering system to be developed very early in the project design phase as well as the format of engineering registers which are used to record information of this type.

Part of the database will comprise the materials to be used in the structure and equipment. Material selection requires a specialist knowledge encompassing the service and fabrication requirements. The design team should include a specialist, although recourse to external consultants is often practised where the subject is on too small a scale to justify an in-house specialist. Acquisition of the correct materials relies on their clear and accurate definitions on the drawings and specifications, so that they can be identified in the material take off activity.

Material selection decisions are ones which should be specifically covered in design reviews because of their specialist nature.

Checking and Approval

Checking is carried out at a number of levels, the most basic of which is that undertaken on individual drawings and calculation sheets. Such checks are usually made by one of the originator's group by someone of equal or superior experience. This is known as a single discipline check, errors can quickly be corrected, when found, before the document is issued for wider circulation. There is some doubt in the case of calculation checking whether this process can be fully effective unless the checker also makes the calculations himself. There is always the risk of missing errors or repeating the error in merely checking by reading through someone else's work. The presentation of hand calculations should be to a format which identifies clearly the subject of the calculations, their status and evidence of checking. The calculation sheets should be bound in volumes with a means of indicating where a particular volume falls in the overall series. Clarity is achieved by spacing out the material, which also gives a benefit of leaving room for the incorporation of minor amendments without having to re-arrange all the page numbers.

In all but the most simple project several engineering disciplines are at work, for example, process, electrical, instrumentation and structural engineering. Although each discipline has to develop its own individual part of the design, it cannot do so in isolation and without due regard for the constraints placed by the other disciplines. The structural designer designs a structure to support the weight of the plant and must, therefore, be advised of the weight and distribution of the equipment. The size of individual items dictates the disposition of structural members. At the same time, the piping designer has to connect items but needs to avoid structural members. Heating and ventilating require large ducting which has to occupy the space which the piping and cable trunking may share. An important form of checking is, therefore, the interdisciplinary check. This type of check carries more importance as the design progresses, for it is then that apparently minor amendments or modifications required by one discipline can have a disproportionate effect on the work of another.

Even though the computer software may have been calibrated, it will still give incorrect answers if the input is wrong. The checking of computer input is, therefore, extremely important, not only from the engineering point of view, but because useless runs cost money and occupy time which others may need to keep to the programme. Checking of the output is an obvious but important activity requiring personnel of some experience who can spot evidence of marginal validity or, less likely with a verified program and checked input, complete rubbish!

The checking of computerised draughting requires skills of a different nature, but again, the checker needs to be able to get into the basis of the design. His task is to ensure that all the necessary lines are drawn, that structural members and joints are rational, that process, electrical, piping and instrumentation lines are correctly presented and that materials are properly identified. An associated task is to reconcile the computer models with the design drawings.

Checking material take offs (MTO) is one of the final checks and one which is very necessary to avoid shortages or delays during construction.

These can all be considered basic quality control actions or "inspections" of the design activity.

Control of Design Work

This function is principally one of an administrative nature. It will have been deduced from the foregoing sections of this chapter that much depends on the recording and transfer of information in the form of written documents or drawings. On the scale of an offshore platform design project, the number of companies and people involved is quite immense and it is quite useless to rely on informal arrangements for transferring the information or updating documents. Systems must, therefore, be used and these systems must be controlled to ensure that they work. Such control must have a reference point, and this has to be established in the form of the documents which represent the status of the work at any time.

In a design office, where manual typing and draughting are practised, the original is usually the master document and any amendments are made on that as a base. For facilitating changes to a drawing a "second" master is sometimes used; this is a reproduceable copy of the original. In offices where word processors and computer draughting are practised, the definition of a master document is more difficult. In this case, the "original" may be thought of as the stored data in the computer memory or disk. On the other hand, this may be seen as a virtual image of the document which cannot be read at will or without the assistance of the electronic equipment. A solution is to produce a paper, or hard, copy of the document which is designated the master copy. Whatever solution is adopted, it must meet the requirement that the master represents the status of the document at a fixed time. Changes extend from the master only.

There is a natural extension from that subject to the security of documents held in computer stores, on disk, tape or in any way which cannot be checked by eye. It also applies to the computer model of the item being designed. The calculations and specifications refer to the model in a specific state; there needs to be a way in which this state is controlled after any stage of amendment so that unauthorised modifications cannot be made. In the ordinary run of work, this can be established by a working procedure and the issue of passwords only to those who are entitled to work on the model.

In a process plant such as that on an offshore platform, electricity is used for almost every operation, and for those which are not, probably for its control. In addition, the services providing a living environment for the crew rely on electricity as do the navigation and communication systems. It is, therefore, necessary to keep a check on the electrical demands, both in total, and in terms of operating demands. For this purpose, an electrical load schedule is prepared and maintained as a measure of the needs of the generation plant and distribution system. This is most important to the effective and safe operation of the installation and must be controlled throughout the design of the plant.

Weight Control

Weight is an important characteristic of an offshore structure. For the substructure, the success of the installation procedure depends on an accurate knowledge of the weight and its distribution; installation also requires an accurate prediction of topsides weight and centre of gravity, whilst the design of an efficient substructure requires early confirmation of the topsides budget weight.

It is important that the responsibilities for weight and centre of gravity estimating are clearly defined. Agreement must be reached at an early stage in the project on such matters as contingencies and load cases for particular operations, dry weight, lifting weight, operating weight etc. An effective weight control system will be based on an initial budget; information supplied by the design offices and vendors will be fed into the system to give a continuous record from which trends can be deduced. Action can then be taken if the budget tolerances are likely to be exceeded. A computerised weight control system will give the quickest response to a given input but even this cannot be effective without up to date and accurate information.

It is, of course, possible to check weigh individual modules and pieces of plant and this is, indeed, done. At the stage at which this is possible, it is too late to make any but the most minor changes to the actual units and limited changes to the lifting gear. It can in no sense be regarded as part of the weight control operation, although it is a confirmation of the effectiveness of the system. Weight control is, therefore, another of the specific issues which must be considered in design reviews.

Technical Interface Control

This activity concerns itself with physical interfaces and not those dealt with in interdisciplinary checks or management communications. Examples are the connections between two modules in terms of structural and system connections, the requirements for the helicopter landing area, crane support structures, drilling rig installation and topsides to substructure interfaces.

The first task is to identify all the interfaces and allocate the responsibilities for their development. The interfaces will be documented using, as far as possible, existing documents; an interface document schedule will record these and their distribution. One or more engineers may be given the sole responsibility for controlling interfaces and it will be apparent that their work will impinge upon the activities of document control and inter-disciplinary checking. Interfaces are not boundaries at which responsibilities are handed over; it is the responsibility of the interface engineer to ensure that connections are made across the interface. An unpenetrated interface is a failure of the interface engineer to achieve this.

Engineering Change Control

This title includes the word "engineering" to distinguish it from the activity which deals with variations which are changes in contracts due to alterations in scope, terms of reference or circumstances. However, it will be understood that engineering changes can be the cause or the result of contractual changes.

Demands for a change can occur at almost any stage in project design. By its nature, there is no reason for instigating change control until a baseline design has been accepted. This baseline must, therefore, be defined on the basis of the master documents referred to earlier. Thereafter, changes are not authorised until the basis and the effect of them has been assessed. This will include the immediate engineering effects, the "knock-on" effects and the cost implications. The latter may not interest the quality assurance surveyor but the other items may well do. What is of importance in quality assurance is that any change must be acknowledged anywhere it may have an effect. Proposals for change must then have an appropriate circulation and opportunity for response which is recorded in a register. The proposal will then be reviewed by a board which may consist of the project manager or project engineer, senior discipline engineers, planning engineer, weight controller, interface enginer and anyone else whose role encompasses potentially affected areas. This board will make a decision to accept the change or not. If it is accepted, the change order procedure will be implemented. Such a procedure will meet the general quality system requirement that design changes are subject to the same controls and authorisations as original design work.

Data Base Control

Data base is a term which has come to include all the sources of technical information available to the engineer. It encompasses written material as well as that held in computer stores. In engineering, much of the data used is derived rather than absolute and it does change with time. For example, fatigue data for welded joints is revised as research reveals more of the influences of variables, such as sea water and material thickness. Most of this type of information is contained in standard codes of practice or specifications. It is important that the latest information is available to the designer; this requires two levels of information control. One is internal which ensures that the latest issue of any document is distributed from the library to all document holders, the second is external whereby the library itself ensures that it is aware of, and acquires latest issues of, the documents in use and any new documents which may be published. This will require a mutual operation between the librarian and the engineer; the latter should be in a position to identify new documents through his routine professional contacts and institutions whilst the librarian receives digests from the standards organisations which he can pass on to the engineer.

The quality assurance department should see that effective mechanisms exist and are in operation. An audit should be able to detect any shortcomings in this area since it can be conducted against the defined project data base. This data base itself must, of course, have been drawn up with the assistance of the librarian. Some data is specific to a project and will be supplied, perhaps, by the client as in the case of the characteristics of gas or oil. Other data, such as soils data, may be provided by specialist contractors. In these cases, the usual document control systems must be able to ensure that any revisions to the documents are conveyed to the users.

Operability, Availability and Maintainability

These three separate but interlinked subjects address the philosophy of the operation of the installation. They are of interest to the designer's quality assurance department in that they represent the objective of the design task. In as much as availability is a function of reliability, it is very much a quality measurement. It is also a function of the maintenance policy which is external to the detail design of the installation although it may have played a part in influencing the conceptual design.

The role of the quality assurance department here is to see that the evaluations are conducted in accordance with the standard procedures for that activity.

However, this subject is dealt with in more detail in chapters 5 and 6.

Safety

Of prime importance are the activities which confirm the intrinsic safety of the installation and the operating methods. These are inherent in much of the plant and the equipment design data base and procedure itself but only part of the facilities and procedures for operational control and personnel evacuation may lie within the design contractor's task. The subject is usually jointly reviewed by the designers and the operators with the assistance of such specalists as are required.

Quality assurance requires that all procedures and specifications relating to safety matters in design have been implemented and that these procedures have inputs from the latest statutory and client requirements.

This matter is dealt with in more detail in chapter 5 as it is part of the certification procedure as certification (at least in North Sea installations) is the method of demonstrating that an installation is "safe".

PROCUREMENT

This term usually means the acquisition of pieces of equipment which are designed, manufactured and sometimes installed and commissioned by the supplier. The equipment may be a standard item such as a valve which is purchased off the shelf to the supplier's specification or a one-off item such as a pressure vessel made to a standard specification.

The basis of the contract with the supplier is somewhat different from that in construction or fabrication work. It relies on the equipment being of a fairly conventional or developed nature and the client accepts it as a complete unit. The design contractor specifies what is required in a document called a requisition. This is, in fact, a detailed specification normally prepared by the design engineer and sent to the client. This is accepted by the client who then issues a purchase order to the supplier accompanied by the data as on the requisition. The supplier may, in fact, already have seen the requisition if it was issued as part of the preliminary enquiry to potential suppliers. The price would then have been based on the information in it.

Vendor Appraisal and Audit

Suppliers of items in response to this type of order are often referred to as vendors; this distinguishes them from contractors. The importance of quality lies here as much as in any other activity and it is the practice to assess vendor's facilities and procedures to see if they have the capacity, skills, experience and personnel to suit the project. The design company, perhaps with the client, may make a visit to the potential vendor's premises to make an assessment, or appraisal. This vendor appraisal may attend to all areas of the facilities but it may be accompanied or followed up by a formal quality system audit. This will be designed to assess the quality programme of the potential vendor. If it meets the designer's or client's requirements then the way is clear for the vendor or be considered suitable for supplying the particular type of product. Minor inadequacies may be pointed out to be rectified by the vendor but any major quality system faults will exclude him from the list of approved vendors for the project.

This is a time consuming and costly activity for all parties and in the UK a number of the oil companies have set up a common quality system assessment organisation, QUASCO, whose appraisal of a vendor will be accepted by the constituent companies. There are other such assessment schemes outside the oil companies, which may perhaps be accepted by them in time, for limited product ranges.

A number of third party certifying bodies (not to be confused with certification authorities) are in existence, or being organised, to provide quality system or product approval of suppliers. These will all operate under the guidelines of the National Accreditation Council for Certifying Bodies (the NACCB) to provide independant verification that the suppliers' quality systems meet the criteria of BS5750 plus any specific product standards. Schemes operated by British Standards Institution, Lloyds' Register Quality Assurance Ltd and, in the case of fire prevention equipment, The Loss Prevention Certification Board, are likely to be particularly relevant to the Offshore Oil and Gas industry.

The vendor approvals given by major public purchasing bodies (eg power utilities, defence ministries, national telecommunications networks) may also be acceptable in certain cases.

In a similar way, The American Petroleum Institute (API) is setting up a scheme to audit and assess vendors against their quality system standard API-Q1 (see chapter 2). The aim is that such approved vendors could be used by clients without the necessity of further auditing.

As with all such approvals, the prudent purchaser will check that the programme which met with approvals is still being implemented satisfactorily before placing an order and that it is relevant in all respects to his needs.

Criticality Ratings

Cost and practicality require that the level of concentration on quality assurance and verification of an item must be adjusted to suit the needs. In this spirit, a simple item whose failure would be of no consequence, will have a very low level of attention, whilst a complex piece of equipment upon which the safety of personnel depends, will justify a high level. This is where the concept of criticality ratings discussed in chapter 2 is of most value. It is now routine to use a criticality rating system in procurement.

However, in spite of this logical approach to assessing the true quality assurance needs, it is found that there is a need for an overriding assessment in some circumstances and this is given where the results of the rating assessment are at odds with the perceived consequences of failure. The factors are agreed between the design engineers and the quality assurance manager. For example, it can be seen that the design philosophy in respect of redundancy will influence this criticality rating for a structural component.

Product Release

In the case of components or fabricated constructions, each item can be released only with the approval of the designated personnel in the inspection procedures. This is an extremely important stage in the progress of the work for it represents the moment at which the title of the goods passes from the manufacturer to the owner, or to another organisation in the chain. There are, therefore, legal implications involved, but in a practical sense it is the moment when the manufacturer can no longer undertake repairs on his own initiative. It is a key milestone in the project plan and must attract the attention of all the personnel from whom authorisation is required on an objective basis. All documents recording the quality control of the item should be collated into a compact form known as a data book which accompanies the item throughout its further installation. It remains as a record demonstrating conformance with the specification. In a project where the design contractor is responsible for the purchasing function or, at least, approval on the client's behalf, this release activity is a signal that the vendor can be paid for the item. This area is developed further in chapter 4.

CONCLUDING REMARKS

In considering all the different tasks and responsibilities in project management, design and procurement (particularly for a large or complex project) we can see that they can all match or relate to the requirements of a formal quality system and, therefore, it is reasonable to assess or audit against the standard quality system criteria.

However, there is more to the subject than that. The complexity of the tasks makes it essential that quality assurance (and assessment of compliance with quality system criteria) is undertaken by people who have some expertise in the business and not just a knowledge of "quality assurance".

Some other points of importance arise. Firstly, the project quality plan must be given great importance as it is the guiding document for all quality related activities in a project and will be the guiding document for quality assurance purposes. Secondly, the communications and control of documentation is a major part of the quality assurance activity. Design, procurement and project management, inasmuch as it produces specifications or drawings, rather than hardware, is particularly reliant (more so than manufacturing) on the control of information flow (not forgetting the flow of information across interfaces). Finally, the documentation of data is, in itself, important to demonstrate the existence of procedures, the correctness of design steps and the achievement of project requirements.

REFERENCES

3.1 Peddie, R. A. *The management of large, high technology projects*. Paper to the Institution of Electrical Engineers, February 1976.
3.2 Wood, K. and Benham, D. J. *Quality Planning for Offshore Production Platform Projects*. European Petroleum Conference, London 1982.

CHAPTER 4

QUALITY ASSURANCE IN CONSTRUCTION AND INSTALLATION

K. P. Bentley

INTRODUCTION AND SUMMARY

The construction phase of an offshore project is, in many ways, the most difficult to control mainly because of the large number of suppliers of all types (who may well be geographically dispersed throughout the world) with the consequent communication and interface problems.

In this chapter we will look at the topic of quality assurance in construction and installation primarily from the point of view of the owner/operator because it is he, after all, who defines the quality assurance requirements of a project and must, therefore, set the pattern. Contractors, subcontractors, fabricators and other suppliers must, therefore, develop quality systems of their own to match these specific requirements of the owner/operator—ie they must react to the owner's requirements. Any appraisal of the managerial arrangements that are considered and implemented to obtain the required technical quality and also to demonstrate convincingly that it has been achieved, must necessarily encompass the very wide range of physical items and assemblies that go into an offshore facility. It can also usefully draw comparisons and distinctions with large undertakings of a similar nature on land which have as their basis many of the same manufacturing facilities and sources of material and equipment supply.

We will therefore look at what is involved in off-shore projects in the construction phase; the features of off-shore work which make it different to land projects and the consequences which therefore arise in terms of the special requirements and considerations that have to be addressed and allowed for in the overall management of the off-shore situation. The nature and scope of the quality assurance arrangements which must then be made to cover these special requirements is then considered with particular regard to the chains of contractual and technical responsibility which are set up in major projects of this kind.

Attention is also drawn by examples to a number of aspects which have proved troublesome in the past or which have needed adjustment of the quality assurance or quality control systems to accommodate them. For however thoroughly and cleverly we anticipate the detailed technical and range of managerial arrangements relating to quality, every major project turns up something unexpected. Flexibility with a quick response is one of the essential features required in a successful project in this field for which a lively and intelligent application of formal quality assurance methods can be especially helpful in identifying problems quickly and promoting the necessary responses.

SCOPE OF THE CONSTRUCTION ACTIVITIES

There is *no* aspect of an off-shore project which cannot, or should not, be subject to the formalised quality management treatment since we look for an effective and efficient performance in every constituent part of an overall project at all stages of the work, with the ultimate objective of providing the service capability and technical integrity required within the intended cost and timescale.

Good communication, good accountability and good recorded evidence that the required quality, in the broadest sense, has been achieved are therefore essential and these are the very conditions or outcomes that a well structured quality management system that is truly relevant to the engineering objectives is capable of promoting.

If we consider the technical scope of a fixed off-shore working platform to be designed, built, installed and commissioned as a single major project involving the use of formalised quality assurance systems then we can see that this scope is very wide indeed.

Construction must embrace large scale thick tubular steel fabrications with complex intersections and material thickness changes for the jacket assembly and with internal and external tank and pipe systems for ballast control in sinking the structure in precisely the right place. Additionally, large area mud mats, pile guides and long welded tubular piles are also needed to complete the fixing to the sea-bed.

On the platform itself are a range of facilities for the drilling of wells with associated storage of equipment, consumables and pipes for oil and gas retrieval and its subsequent processing, for the generation of power and its distribution necessary for the functioning and the working of the platform and its equipment, to provide accommodation and administration quarters for the resident workforce with communications by radio and arrangements, for transportation and handling both by helicopter and by sea with fixed cranage for lifting cargos on board, and finally all life support systems and safety features including life-boats and fire-fighting systems.

These necessarily involve many items of specialised plant, either requiring subcontract construction specific to the project (such as pressure vessels for process plant, diving decompression facilities and conductor pipes) or proprietary items to be bought in as finished components from specialist manufacturers. These latter are often most difficult to control from the quality assurance point of view and may include items such as mud pumps, valves for process plant, electric motors and power distribution equipment, fire-fighting equipment, life-boats, cranes, instruments and electronic control or communications gear, furnishings and kitchen equipment for the living quarters.

In addition, the isolation and often remoteness of off-shore intallations require reliable support and supply services to be established and in the operational phase an associated network of pipeline and storage installations on the sea-bed nearby enabling efficient and safe removal of the product, either to land by the pipelines or to surface ships will also be necessary.

This in outline describes the scope of an off-shore installation.

COMPARISON WITH LAND-BASED PROJECTS

It is appropriate to make comparisons with large land-based projects such as for power station construction or for the erection of a large bridge or oil refinery or chemical plants for the simple reason that formal QA has been used in such projects in the construction phase for many years. As for a power station or an oil refinery on land, the construction of an offshore platform involves both the fabrication and erection of structural steel work and the building of pressure containing plant for the processes involved. It also shares the requirement for a substantial proportion of its fabric to be manufacturers' proprietary items for particular specialist duties. In both these respects the quality assurance involvement may have many similarities; material selection and procurement, fabrication techniques, inspection and test methods, for example, will be major areas for QA activity in activities such as building the steelwork or vessels and pipework. The monitoring of procurement activities with special regard to the accurate specification of requirements, the early evaluation of prospective suppliers' quality capabilities, input to the appraisal of bids and tenders and inspection surveillance during manufacture are essential for both in this area and for the manufacture of proprietary equipment.

Clearly all quality assurance activities can influence and, indeed, are often required to contribute to, the contracted terms established between parties working in major projects. This is true of both an offshore and a land-based major construction project. A further, and significant (in quality terms), similarity to be found between them is in the multiple layering of contractors and suppliers each having to specify requirements and communicate accurately with their subsidiary contractors or suppliers on all questions of technical requirement, inspection and acceptance standards, concessions and technical queries.

The quality management system set up for the entire project must obviously operate successfully in contracted depth and, therefore, liaise effectively with parties at all levels since it is the requirements of the ultimate user or operator of the installation, be it an oil company or an electricity generating board, that must be safeguarded and realised. This point, of course, was discussed in earlier chapters and it is this interest which defines to a great extent the approach to quality assurance.

The more complex that contractual chains are allowed to become, the greater the difficulty of achieving effective and efficient communications which increases the danger of technical problems remaining unsolved for too long and places extra duties on the project's quality management system. In the previous chapter the complexity of contractual chains was indicated and its importance in respect of project management and design. The same reasoning applies to the construction and installation phases. For example, there have been instances of technical queries being raised by a component fabricator and never receiving a *formal* contractual decision or answer from the ultimate client. In such a case the fabricator has to act on the provisional solutions unofficially conveyed by the agents of the insuring or certifying bodies involved or just rely on his own initiative and skill. These situations arise primarily because too many contracted parties are involved for speedy decision making.

As a guiding principle therefore, owners should aim to make all contractual chains as simple and short as possible and before any additional intervening contrator is allowed to become involved the user should ask himself very carefully what engineering benefit and genuine technical contribution can be provided by an extra contractor in the chain. This should be the aim. Unfortunately it is not always achievable because the very large range of expertise needed to complete a large construction project normally require an equally large range of organisation or groups to be involved.

This is particularly the case when a project is being run by a user (often one of the smaller users) whose technical expertise has been in areas other than construction. There is often then a tendency to complicate the management of a major project, perhaps unnecessarily, by involving major large contractors to effectively manage the entire construction project.

In such a situation the ultimate client can then lose direct contractual contact with the component fabricators and suppliers wherein reside, usually, the greatest specialist engineering knowledge and technical skills. In such a situation it is probably the fact that the capability for carrying massive financial responsibility and investment has been allowed to override the true capabilities for engineering and technical achievement. An interesting trend within the long established and major oil companies, who have in themselves an accumulated wealth of technical experience in both onshore and offshore construction, is for "in house" project groups to be set up, jointly comprising an overall project team, so that direct contracting between groups and principal fabricators and suppliers can be used. Whatever the main reason for this trend, it should certainly ease the communication problems inherent in the quality assurance of major offshore construction projects.

In this sense, offshore construction projects are moving in their management structures closer to the systems used and well established (in the UK) for fossil fuel power stations on land. The relevant UK public utility (the Central Electricity Generating Board) has, over the years, developed the technical and quality management capability to operate in this way and this is the ideal way to operate. It does require, however, well-established engineering and managerial teams and traditions which may be harder to achieve in the more volatile offshore oil and gas industry.

So much for the similarities between offshore and land-based major projects; now we can consider the major differences which arise as follows:

1. Because of the many different functional facilities required on an offshore platform and because of the particular problems of transportation to an offshore site with the associated need to complete installation rapidly it is customary for the top-sides of an offshore platform to be designed and built in modular form. That is to say, as large boxes of structural steelwork into which are fitted the required facilities, complete and ready for operation.

For a large platform, the total combined dry weight of the modules could amount to over 11000 tons and comprise over 20 separate modules. Typically, these could comprise two for well-head operation, six for drilling activities, eight for production (including the generation of power), three for accommodation and a helicopter deck on top. Module construction involves a number of discrete stages and planning and contracting of their construction and of the entire project is usually based on them, namely:

(1) Fabricating of structural steel work (that is, making the box)

(2) Fitting into it all pipe-work

(3) Installing all other plant and equipment

(4) Architectural and finishing work such as insulation, fire proofing, painting, vapour seal membranes

(5) Furnishing (in the case of accommodation, catering and administration office areas)

It is usually necessary to contract different specialist subcontractors or at least involve different teams of tradesmen and practical skills to accomplish these various stages.

When modules are completed in all the above stages they must be loaded out on to barges to which they must be fastened securely and taken to the jacket already in place and piled to the sea bed. Installation involves lifting them by crane on to their intended locations and welding the beams and ends of the box structure into the platform structure as a whole. The need for dimensional accuracy is obvious and when the structure is complete, the connection of all piping systems and electrical circuits (ie the hook-up phase) can proceed. These two stages again require different skills and often different contractors.

It is also at these installation and hook-up stages that the effectiveness or otherwise of the QA system as a whole is especially revealed when modules that do not fit together within fabricating tolerances or pipe systems which fail to match their corresponding parts become obvious. Module structural steel work must be carefully designed and fabricated to withstand the transportation and lifting loads which they will experience both on the barge and when they are set on top of the jacket structure. The steel work may to some extent then become redundant, since working loads *in situ* may be low in comparison with those experienced at load out and installation.

Close collaboration with a company specialising in transportation of large items such as jackets and modules and equipped with the necessary barges and tugs is clearly essential from an early stage in the project. The actual platform site, the positions of lifting lugs, the size and number of modules and weight limitations can all be influenced by the available means of transport and movement of the items has to be included in the early design stages.

Not only should the QA department in the user/operator organisation help to promote this liaison but a close involvement with the transport company is necessary regarding the rules to be applied in the fabrication and welding of sea fastenings since the transport company will inevitably insist on its own welding capability being used for attaching the modules etc to its own barges. The welding procedures to be used must therefore be compatible with the fabrication specification for the module structural steel-work—ie a potential interface problem.

2. In offshore construction projects there is a degree of urgency not usually encountered in similar land-based projects. The installation programme, often involving transportation of jackets and modules from several different yards often in different parts of the world and planned for what is often a restrictive weather window at the site, demands that each item must be ready at the correct time and by ready we mean of proven, correct and contractually accepted quality with supporting documents and certification

all completed (this documentation aspect is very important and will be discussed further in the next chapter).

Failure to meet the load-out deadline in every respect costs heavily in projects of such enormous value and any subsequently lost production days through late completion are seen as direct financial losses.

3. We will see (chapter 7) that consequential costs are significant costs even on land-based projects. These costs can be multiplied many times for an offshore project.

In looking at these major differences, we see that they are reflected in a number of special features of construction which related to the QA approach:

Firstly, because we stand the jacket with its platform on to the sea bed it is essential that a knowledge of the local geology be obtained. This enables a prediction of the piling characteristics to be made and most important it imposes and sets a weight limit for the platform as a whole. This weight limit must be regarded and held at all design stages and throughout all stages of construction of the modules and during the accumulation of all the pipework and equipment to go inside them. In this respect the quality control system is particularly valuable in monitoring the weight values reached at all stages of module construction.

Secondly, the environment is unquestionably more hostile than the situation on land. The sea, waves and winds ensure that the platform is going to be subjected to not only high static loads but also significantly fluctuating loads, involving complex fatigue stresses. Sea water corrosion conditions also apply and the low temperatures to be encountered in an environment such as the North Sea or the Arctic place particular demands on the materials used and on the skill of fabricators especially with regard to welding.

Indeed to the metallurgist or material scientist the onshore/offshore comparison is between the selection and control of materials designed for use at elevated temperatures in land-based process or power generation plant and materials designed to have particular low temperature properties for offshore structures. A frequent lack of ease of fabrication especially with regard to weldability is often associated with both types of material and fields of application.

Thirdly, the working processes involved place further restricting demands on the choice of materials used, especially for pipe work where crude oil and gas may present a whole range of corrosion and erosion problems associated, for example, with hydrogen sulphide, carbon dioxide, phthalic acid and suspended solids, etc. Added to this are the fire hazards created by handling highly inflammable gases and hydrocarbons in a closed and isolated environment where fire fighting and evacuation are more difficult than on land.

Fourthly, unique to an offshore platform is its multiplicity of purposes; not only is it a working place as a chemical process plant but it has also a power station and an airport within its framework. Add to this the fact that resident operating staff must live there for prolonged periods and it has many attributes of a ship at sea.

Like a ship, it is a self-contained living and working unit which can sustain itself for limited periods, but which is not totally independent and must depend for its continued existence and intended function on systematically planned and organised supply services.

Fifthly, related to this, because a platform is necessarily big and technically complex, its construction involves a very large number of different kinds of suppliers of materials, parts and services and of subcontractors to perform the fabrication and erection (each providing their own specific skills and technical capabilities). The task of ensuring that every one of them is made aware of the quality policies for which evidence is to be sought places particular demands on the quality management system as it permeates from the user or operator at the top of the contractual chain to the especially wide disparity of subsidiary parties beneath the user/operator.

Sixthly, although all major projects take a long time from inception to completion, the construction of offshore installations is probably unique in having time scales which are comparable to the rate at which technical developments occur and become available as innovations relevant to the work and which could be introduced during a project life.

For example, developments in steel making and consequent improvements in the engineering properties of structural steels have been significant in this repect so that steels specified for one major offshore project are unlikely to have been specified in exactly the same way for the next (the story is similar for welding consumables). Furthermore, in many parts of the world, the structure of the manufacturing industries who can provide all the component parts and the services required has altered to such a degree that fabricators, suppliers and service companies who have been involved in offshore work may change ownership or disappear. They come and go not only between projects but sometimes during the life of major offshore construction projects.

For a great many reasons, therefore, every installation is unique contractually and in a technical sense. The consequence of these broad considerations lead to a number of aspects and activities to which control must be applied on behalf of the ultimate client at all project levels and in this, the use of formalised quality management systems must play a vital part.

CONTRACTUAL APPLICATION OF QA

We have seen from the previous section and, indeed, previous chapters how quality management is appropriate and necessary at all the contractual levels of an offshore project.

Although the concept must necessarily spring from the interests of the ultimate client (the installation is after all intended for the user/operator at the end of the day) it is worth considering also how other parties involved in construction and installation, such as the principal fabricators and suppliers of materials, components and services, may see and react to having formalised quality management systems established to which they must contractually comply.

Offshore construction projects are characterised by high investment and acute safety considerations so for the owner/user/operator the prevention of mistakes of any kind either during construction or in subsequent service must be a prime target for his *own* QA activity and for that of all his contractors, subcontractors and suppliers.

We must recognise that any activity if not performed correctly or any material or component that may be less than ideal can lead to unwanted situations either sooner in the construction yards or later when oil production is in full swing or worse when an emergency situation arises during operations. In any such event lives, time, money or production capability may be lost. The owner's QA efforts should therefore not only promote the final technical quality but contribute to efficient project operation and progress by minimising unwelcome delays and extra costs.

The owner/operator has also to keep in mind that before he can put the installation to work and create a profitable product it must have been given all the necessary approvals and associated certification from the Statutory or Certification Authorities appointed by the Government in whose national waters the platform stands. The owner's QA system must therefore systematically generate all the necessary documentation relating to design, construction, installation etc for these authorities so that the plant can be permitted to perform in production as intended.

This must necessarily involve inclusion of documents and quality control records from the many contractors and suppliers involved and is one of the prime reasons why the QA department must relate directly to all other contracted parties. This certification and documentation aspect, of course, is discussed in more detail in the next chapter.

In addition, the quality system should ensure that detailed technical records for the owner's later use have been provided relating the materials used, the design, methods of construction, quality levels actually achieved including repairs or changes made and indicating any technical concessions or amendments to specification that were allowed. This enables both modifications to the structure or plant to be made in later years and routine maintenance during its service life to be undertaken on a sound technical basis.

Similarly fabricators require a smooth flow of work at site and in their construction shops to promote production efficiency and they look for a ready acceptance of their products by owners and regulatory authorities.

The economic advantages of getting it right first time in every kind of project cannot be overemphasised. "A fundamental truth of all engineering and manufacturing enterprises is that whenever events turn out less than ideally someone always loses either in money or reputation and often both." The significance of major structural or plant failures at an offshore installation is easy to interpret in these terms as a direct loss of production time or equipment usage for the owner and everyone in the manufacturing chain, whether main contractor, fabricator or material supplier, may have back charges or a damaged reputation to stand. The future implications regarding human safety are obvious.

However, it is equally true in the technical details of the fabrication shop, ie the place where quality is actually created in the product. For instance the replacement of wrongly used material, rectification of a machining error or the repair of a defective weld represents a loss of manufacturing resource for which a sum of lost money, often surprisingly large, can be calculated. These are the losses which are collectively known as quality losses and are discussed in chapter 7.

The avoidance of these potential losses in financial return, image or manufacturing resource is the principal aim of the fabricator and his means of achieving this trouble free product acceptance is through the efficient and effective quality control of his manufacturing activities.

The specific quality controlling activities, checks, tests and records that a fabricator or supplier is asked to perform by the user, or more often his representative, have not infrequently led to some friction between the user and the supplier especially in matters of welded construction.

The supplier should know from his own practical experience and technical capabilities just what are those human mistakes, oversights, technical errors and sometimes material idiosyncrasies against which quality control measurements and monitoring must, protect him. On the other hand, the user's viewpoint is more general and may be more doctrinal and often seemingly less relevant to the practical reality of technical details in structural hardware.

This contrasting viewpoint is often exacerbated by the presence of a major main contractor acting between the ultimate user and the fabricator and situations of outright confrontation have occurred between fabricators and main contractors and indeed between main contractors and their ultimate client, the user. Generally when fabricators or suppliers with a superior knowledge of, say, welded construction deal directly with the intended operators who know better than anyone what it is they want to have, then matters can run more smoothly and more successfully in technical terms. Though this does depend, as was stated earlier, on the owner/operator having sufficient technical resources to effectively deal with suppliers in this manner. Loss of contractual initiative by the user/operator in appointing a large main contractor has, in the past, often led to a proliferation of technical requirements expressed through a multitude of extra project documents and specifications ostensibly written from scratch for the new project but in reality based squarely on previously documented practice.

The tendency has therefore been for technical specifications to become increasingly rigorous with regard to materials and construction practices and the tests and checks to be performed. This has not always been to the ultimate technical improvement of the final platform's true engineering status and to the fabricator performance. He is often faced with an ever increasing number of weld procedure tests to perform, many of which are exact repeats of his last major offshore contract and which he knows are unnecessary in terms of his own welding management and often technically irrelevant to proving the engineering integrity of the final structure. The cost and inconvenience of this aspect alone has been well documented (ref. 4.1). In this particular context, it is noticeable that a direct contractual relationship between fabricator and ultimate user has very often enabled previously qualified welding procedures to be approved and used without re-testing, thereby reducing the extra effort involved for much work. This is hardly surprising since contractual liaison of this kind has been working successfully and agreeably for years in, for example, the construction of power stations on land. This indicates the potential for minimising the cost of quality assurance without reducing the level of assurance *provided*, as we have stated, that the owner/operator has sufficient technical resources to take this route.

Quality assurance and its implementation by quality control measurements and documented records are therefore a significant ingredient of the contractual terms and relationship of any offshore construction project. The activities to be undertaken and the safeguards to be built in have to be clearly specified and defined *contractually* at the outset. It follows, therefore, that the owner/operator should ensure that his *own* QA department is established early in the project and in time for its work to be effectively incorporated into the early placing of contracts, for example for long lead items or for the purchase of the bulk materials to be used and issued free to the principal fabricators when appointed.

This is a specific need for the construction phase of a project on top of the other planning and control needs and is one which is not always given due attention. For example, experience has shown that the majority of technical problems which arise during the life of a major offshore construction project have as their origin either some deficiency in a technical specification (such as a shortening in its scope, some ambiguity in its interpretation or even a direct conflict of requirements) or a fault in the system for placing procurement contracts, often because of insufficient detail in the technical requirements to be achieved or in the selection of an unsuitable contractor or supplier.

We can therefore define a list of specific actions or aspects which the user/operator should address when embarking on a major offshore construction project and which must be "matched" by his contractors and suppliers.

EARLY QA ACTIONS FOR THE OWNER/OPERATOR

If QA in offshore construction work has sometimes been less successful than had been hoped for then it is probably because owners have often tended to act too late in the life of the project for QA to be effective and then possibly the wrong kind of effort and activity was made. The so called QA activity has then been a detailed and rigorous inspection and record generating activity rather than a true quality assurance activity.

The following list of specific actions (which have been stated before (ref. 4.2) but are elaborated on here) should be undertaken before embarking on a major offshore fabrication:

1. *Define the quality policy for the project*—This should cater for all statutory and safety requirements as well as structural engineering and process system functions.

2. *Establish a QA department and team early in the project*—The writing of specifications and placing of orders and sub-contracts occur early in the project time span and the QA department must be established in time to contribute effectively to them both.

3. *Ensure that the QA department has the authority necessary to fulfil the user's policy*—Declare this formally to all parties in the project, stating the QA department's terms of reference and defining the relationship of the QA department to other departments in the user's organisation. This might mean a "project specific" QA group rather than a centralised corporate QA group, depending on the way the company is organised.

4. *Ensure good liaison and effective communications between the QA and other departments*—The QA department must be readily available to other departments, especially those involved in materials, design and construction methods etc, and have the means for rapid liaison with then in response to

events. Proper links must therefore be established at all levels in the user's project organisation. In this regard other technical departments should understand clearly their own inherent responsibility towards quality in their day to day engineering work. They must understand that a QA department or QA project group does not create quality in the product—it is only the manufacturers and fabricators who do that—but that QA will monitor and record the achievement of quality in both the product as it is made and also indirectly in the related design and project liaison work performed by other departments. Access to the quality department must be easy and its physical location is important in this respect and should be close to the engineering or design teams preferably in the user's central offices (or the project headquarters).

The appointment of an outside organisation to act as a QA department for the owner, possibly as part of a broader management services package, is formula for unwieldy communication and slow, ineffectual QA response during contracts since it effectively introduces a further layer in the chain of communications. In the early stages of a major construction project, the project management team will be primarily concerned with the ultimate aims and hoped for results of the project in terms of its ultimate market and commercial success. For example, it will be concerned with anticipated yields from the offshore facility, the quality, cleanliness and value of the expected crude product, site geology, the financing of the entire venture and its parts, time scales and forward marketing plans and hence to the logistics of construction and planning for installation and commissioning of the platform with questions that then arise concerning available manpower, construction sites and their capacities, manufacturing facilities and seasonal weather patterns in the area etc.

We have already seen how some actions primarily and specifically related to the QA group must be dealt with early on, but there are also a number of actions which we will now list and which although having some QA implications remain primarily project-wide management actions. These relate to how the project is to be conducted and controlled and involve a QA consideration because they are essentially to do with information and communication and sometimes with the public image and political implications.

5. *Ensure that technical specifications accurately reflect the owner's technical policy or requirements*—If specifications are too lax then some degree of contractual control is lost. It is extremely important that the levels of quality required are not overstated. Specifications should indicate as precisely as possible just how good a structure, product or weld really needs to be. They should be written with true regard to the likely freedom from defects, weld defects and tolerances and properties to be achieved, for instance, that is reasonable to be expected from the manufacturing facilities available.

Specifications which pitch the required quality levels too high can lead to difficulties both technical and contractual during the fabrication stages. This may have any of three consequences:

(i) A large number of unnecessary repairs costing time and money and often with an overall technical deterioration (eg the introduction of further residual stresses and metallurgical damage in a weld being overlooked in the quest for a clear radiograph).

(ii) The temptation for staff in all parties to bend the rules to keep the work moving, eg the designation of a "lack of fusion" as a "slag inclusion" in a weld (defects which are sometimes difficult to distinguish), by an inspection technician when pressure vessel standards (which excluce all lack of weld fusion) are applied to module structural steelwork. We can sympathise with but not condone such actions.

(iii) A string of concessions having to be given by the owner to the fabricator and sometimes variations to the ruling project specification having to be made, approved and distributed throughout the project organisation. Clauses in specifications cannot be waived at a nod from the owner or his representative without recourse to a formal, documented variation procedure because ultimately it is against the specification as written (and *formally* amended) that the appropriate regulatory authority will give a "consent" or a certificate to operate.

In this latter respect, it is worth remembering that the status and authority of a supplier's quality manager, diligently striving to maintain standards of workmanship possibly in the face of conflicting production pressures, may be seriously undermined if a stream of concessions is seen to come from a client.

The owner's manufacturing specification should therefore define quality standards as far as is practicable to reflect the severity of engineering duty that applies, eg avoid simplifying the standards or specifications so that in the extreme, tubular node weld requirements are equally applicable to simple butt welds in secondary and minor bracing. In other words, the owner must bear in mind the twin requirements of a product specification (quality standard and quality assurance standard) as we have stated in earlier chapters.

6. *Adopt a flexible strategy for the application of specifications*—As a corollary to the above the user's strategy for implementing his specifications should allow some flexibility in their rigorous application so that minor imperfections which are truly insignificant on a localised engineering basis and to the integrity of the structure as a whole can be left alone. This must be declared to all departments in the user's organisation, where emphasis should be to detect and measure imperfections, assess them correctly by sound engineering analysis and only repair or rectify those which are truly unacceptable.

The guiding principle should be to find true justification for the acceptance of welds, materials or equipment put under question and not to take delight in demanding repairs or replacements on principle. The mere act of attempting a repair may impair the quality of a structure in reality, as has been well documented (eg ref. 4.3).

7. *Adopt a policy for the control of material*—In most large offshore projects the materials for the structure, both tubular and plate, are purchased by the user and issued "free" to the selected fabricators.

Provision must therefore be made by the project team for the storage, movement and issue of large quantities of steel plate and tubes. In the UK and in European based projects, storage is often provided at the actual construction yards themselves, but in more remote parts of the world there is often a need for a special storage facility to be established. This will require its own security arrangements, involvement of the QA department in receipt inspection, regular condition monitoring and also an established means of transportation to the working locations.

When a storage depot of this kind is established within supporting distance of a fabricator, then it must also be expected to receive, store, inspect and issue large proprietary items of equipment like mud pumps, slug-catchers, generators and items of process plant which must wait somewhere until the construction yards are ready to receive and install them.

8. *Set up arrangements for support services*—The need for background technical support and information is also essential to any large construction project and an efficient and comprehensive library facility must be established, if it does not already exist, to which technical and QA staff can refer. This should include all the current project ruling documents as they accumulate during the project with all relevant national codes, standard specifications, recommendations and guidance documents etc, together with the important published technical literature and journals which relate to offshore construction work.

For technical advice and supporting research or development work the project team may also need to establish working liaisons with appropriate consultancy services, research institutes, university departments and sometimes the research departments of major or specialist manufacturing organisations. Areas of technical doubt may well arise during the life of the project and there should be a facility established within the user organisation for the rapid involvement of sources such as these.

9. *Define organisational and training needs*—The identification of relevant professional and trade skills that will be required as part of the user organisation in progressing the project as a whole is an essential activity before the project manager can set up the relevant technical group within his organisation.

The QA manager similarly has the early task of setting up his own departmental team and there is an early opportunity here for the two managers to establish a good collaborative working relationship in which they can give each other substantial mutual help. There is often a need for specialist training facilities to be set up for personnel employed in the project.

Established training courses in, for instance, welding, non-destructive testing, and paint inspection leading to a recognised and certificated qualification are already well-established, and the project management, with the QA department's involvement only have to agree and document the qualifications and experience they will require as a minimum before employing such people.

Also, many QA staff involved in the project, who may conceivably be required to go offshore during the installation, piling, pipe-laying, module installation and hook-up stages etc, must be trained in the necessary survival procedures appropriate to the statutory national requirements applicable to the platform site. In some parts of the world, such courses already exist, but the user may need to set up his own training course locally to cover his own staff for a capability in first-aid, fire-fighting and survival after ditching from a helicopter. When prospective staff apply for jobs with the user, it is not a trivial question to ask if they can actually swim!

ESTABLISHING THE USER'S QUALITY DEPARTMENT (OR GROUP)

The Quality Manager

The key person in a newly established QA group for a project is, of course, the quality manager as he must liaise very closely with a large number of other groups (see Figure 4.1 for example).

He needs to be well versed and experienced in the use of formal quality management systems and should preferably have held a quality management post in an appropriate manufacturing or fabricating organisation. This will ensure that he knows what life is like in working to formalised codes and specifications where documented accountability is required. Especially valuable is experience in liaising and dealing with representatives of inspection or certifying authorities. He will also be able to realise the true implications and practical significance of particular requirements, tests or inspections that might be imposed and he should furthermore be aware of any countermoves or avoidance tactics which a manufacturer or fabricator may be tempted to try. This needs to be a truly professional appointment and background should be technical in a science or engineering subject related to construction work. Many metallugists, mechanical and process engineers and some welding engineers find a successful life as quality managers in the offshore industry, as do some chemists and electrical engineers.

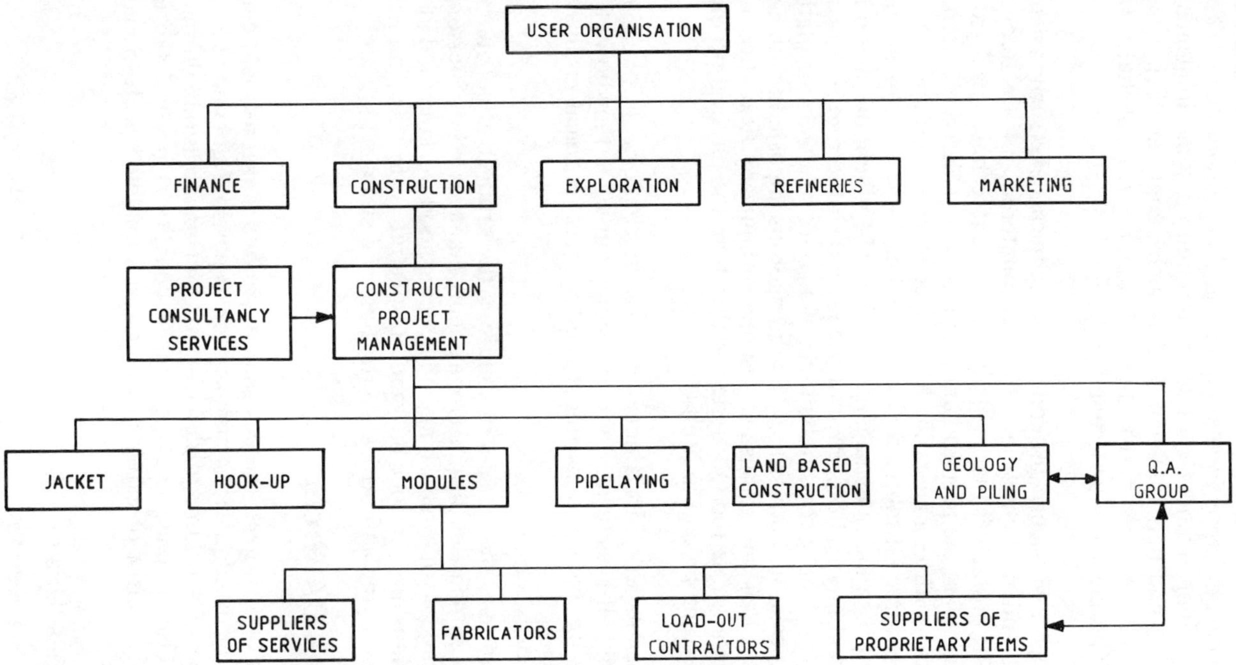

Fig. 4.1 Typical construction interfaces

There was a time when quality manager's jobs were regarded by some company top executives as a suitable repository for old, unwanted, spare or rather second-rate staff thereby avoiding redundancies, but at the same time keeping clients happy by making the labelled appointment. The job itself was appealing to these staff since it promised some influence and authority which had perhaps been denied such people all their working lives, but it led to some narrow-minded and negative attitudes being established. This clearly is unacceptable.

Quality managers with a background in non-technical fields such as marketing, accounts and personnel have been encountered but are frequently unsuitable. The problem is that, without the relevant technical background, realistic discussion of the practical implications of considered requirements cannot be held with engineers in the project technical groups, nor with potential or contracted fabricators and suppliers, however open-minded and intelligent the quality managers might be. They therefore find it hard to keep the real physical objective of the project at the forefront of their considerations (ie the successful building of a good, safe working platform in time and at cost) and tend to rely on the rigorous imposition with a near religious fervour of an established system, without any flexibility, to all situations that arise, ie the QA system rather than the physical product is allowed to become an end in itself.

The quality manager's job is never an easy or comfortable one. When things go badly all eyes turn to him accusingly but when they go well it is generally the production departments and project technical groups who collect the credit.

In truth, of course, the success or failure is the product of collaborative effort and this emphasises the need for a quality manager to be of personable disposition, capable of establishing and keeping good working (and indeed social) relationships, respected and respectful with his immediate colleagues, certifying authorities, contractors and suppliers.

Staffing the QA Group

At the outset of a project, the designated quality manager should choose and establish his headquarters team as soon as possible and one at least of them must be of a calibre to deputise for him on occasions during the project life. Usually collectively known as "quality engineers" the team might typically consist of experienced and well qualified professionals in fields such as:

Engineering design

Welding engineering

Metallurgy

Documentation

Contracting

Instrumentation and electrical

Equipment manufacture

The same engineer may well cover metallurgy and welding engineering and sometimes maintenance and corrosion engineers are included in the group.

They will all need to liaise with project technical groups appropriate to their skills and the documentation man will have a particularly important liaison with the regulatory or certifying authority.

The contracts man should not only co-ordinate the surveillance by the department as a whole of all project sub-contracts placed, but shall be instrumental in drawing up contracts for the department itself—notably in the appointment of individuals to resident site surveillance teams and of service companies for major subcontracted quality control services (eg for weld or paint inspections, special training courses or investigatory work).

The team of quality engineers will require administrative support in the filing and storage of all data and records relating to contractors and suppliers to the various site activities and problems arising and records of quality achievement which are presented for certifying authority approvals and for the final documented dossier relating to the project as a whole for achival use.

Supporting staff will also produce control documents for formal communications on behalf of the QA department and for this should be equipped with full telex and telefax facilities for prompt liaison with site surveillance teams.

QA engineers at the QA department headquarters will be especially active in vendor assessment and specification writing in the early pre-construction stages of the project and the need for establishing site QA surveillance teams to monitor the actual construction work will only become necessary when the construction sites are about to commence work. It is therefore often possible to transfer and relocate some QA engineers from the QA department at head office to the sites as the emphasis of work changes. This can be especially beneficial since those who originally wrote the specifications or control procedures then find themselves applying and using them in the field.

Similarly when construction sites complete their modules or jackets etc., the need arises for offshore teams to be established for the jacket installation and piling, module installation and hook-up stages and members of site surveillance teams can be selected and transferred for round the clock shift surveillance working at the offshore site from attendant barges or similarly on to pipe-laying barges.

The QA manager has a unique opportunity to have studied the working performance of his site staff and especially how they relate to their corresponding project line staff. Personal relationships become especially important in the confined space of the offshore working and team members must be chosen with care for this phase of construction. Arrangements must also be made for them to be equipped with any necessary equipment or clothing for the offshore situation and, as mentioned earlier, they must all have successfully completed a survival course.

SETTING UP AND STAFFING THE QA GROUP FOR INSTALLATION AND HOOK-UP

The application of formal QA to the offshore site or the pipe laying barge is exactly similar to the onshore situation, but because space and especially accommodation are inevitably very limited, the numbers of staff involved must be a minimum.

A multiplicity of skills is therefore looked for in each member of the QA surveillance team. The leader and organiser of shift personnel etc. will have prime liaison with the project line team and will probably also have shown himself capable of meticulous record keeping so that the QC records are clear, complete and always up to date. His inspectors ideally will also have several strings to their bows—an NDT man also covering paint inspection perhaps, and a welding inspector also covering general visual and dimensional checking.

The offshore site, at this stage, will be a self-contained unit regarding the planning, execution and approval for accceptance of the work and although constant reference to the departments or groups onshore is made and, certainly, all problems and queries referred back, the personnel for offshore working are selected to be entrusted with as much on the spot decision making as possible.

In establishing the procedural instructions for the offshore work (ie whilst the onshore yard construction is still underway) the QA manager should make quite sure that the limits of technical authority are precisely defined for the senior project and QA surveillance people when offshore and that they are in no doubt at all as to the kind of circumstances which require authoritative decisions from the shore. Routine progress reporting on a regular basis should already have been defined by him in the procedural documents.

If new working technical procedures are needed for the offshore site operations (for instance, some additional welded attachments may be found necessary during the piling stage), then they would still be subject to formal project management and QA approval from the user's headquarters onshore in the usual way. The use of Telefax communication for rapid turn round of matters of this kind is invaluable but it must always be remembered that any such procedure must be generated in a formal way suitable for inclusion in the final dossier of project certification bearing all necessary approval stamps from the certifying authority.

Very often site engineers and their QA counterparts are changed perhaps every two weeks, so the manning for offshore working must take this into account and provide alternates for each key position.

The actual handover routine when site staff are changed needs to be preplanned carefully so that the engineer being relieved has a full and detailed situation statement for his colleague to pick up, and there should be time during the overlap at site and helicopter turn round time for the two engineers to look over all the work situations and equipment etc. together. This interchange provides an important liaison with the staff at the onshore base and the engineer who has just come onshore, whether project or QA, should expect to spend many hours de-briefing on the offshore state of progress with his department or group head and colleagues.

Nevertheless, some auditing of the offshore site QA activities should be arranged on a regular basis by the quality manager although it is primarily aspects of quality control and associated results and records that should receive the emphasis of this attention.

In this respect the need to maintain all the established methods of formally reporting QC activities and records, ie on standard forms together with the actual inspection or measurement procedures used, must be emphasised, because the QC methods used and the criteria of acceptance are *still* as defined in the ruling specifications and should not differ from the onshore construction yard situation just because it is offshore.

This point of conforming to procedures is even more important to emphasise in the case of the pipe-laying barge. The pipe preparation, welding, inspection and NDT, decision on acceptability and, if necessary, repair and clearance followed by coating and launching down to the sea bed represent a detailed technical sequence which *cannot* be reversed. Therefore, acceptance decisions as the barge progresses must be correct and must be recorded and documented to the last detail. Precise weather conditions at the time when the welded pipe made its descent to the sea bed alter the stresses experienced by each weld in the pipe at that critical moment. If later examination of radiographs of pipe-welds reveals defects of a doubtful size or nature, then subsequent engineering analysis of their significance will require all available information.

Having established the QA department (or QA project group) in the user organisation we can now consider the methods that should be adopted by this quality department and the specific actions necessary to set up the quality management arrangements (and the matching arrangements in the supplier organisations) required.

PRIMARY AIMS OF THE USER'S QUALITY ASSURANCE GROUP

The quality manager in the user/owner organisation must actively promote and stimulate the collaborative liaison of his team with other technical departments alongside in the project both in his own organisation and in the contractor and supplier organisations. He should "set the scene" as it were for the project QA effort. This is an essential prerequisite to any documentary work that must be undertaken. Similarly, the QA department must also from the start establish good working relationships with other external organisations such as the regulatory authorities and any design consultancy services or inspectorates who may be involved. This external liaison is very necessary because views must be exchanged during the specification writing stage (ie planning stage) in which the user's requirements are defined.

To assist in and formalise many of these relationships the quality manager should, for example, ensure that standard project sheets are established—some of their own making—for communication both within and external to the user organisation. This is a project wide requirement and many such standard documents will be generated by those establishing the project management system to which the QA function is complementary. This we have seen in the previous chapter.

The major method of formal liaison however is in the writing of procedural documents by which the working rules to which the project is to be managed can be established. Those dealing with matters of inspection surveillance, methods of testing and rules for documentation control and liaison (ie the management procedures) can (and probably should) be written by the QA department itself but others of technical definition (the technical procedures) such as steel specifications, structural fabrication specifications and component specifications must be the responsibility of the relevant technical departments carrying prime project responsibility for these aspects. It is necessary, though, for the quality assurance department to ensure that these technical documents are complete, controlled and are properly vetted and authorised.

Because offshore installation projects are necessarily large and complex, involving a large number of contractors, the introduction of a quality management system to the many different parties, some in conflict, some in competition, *must* be done by means of these formalised documents. The quality department in the user organisation must be instrumental in establishing these at an early stage.

The documented controls or procedures must convey instructions and clear indications of requirements to all the parties involved or who will take part and there must be clear guidelines established to determine the ways in which inspections and the records of inspection with certification or with signed and dated inspection reports must be made. If, for no other reason, the safety aspects of an offshore construction and the statutory regulations which apply to its occupation and use require documented evidence of conformity with an approved technical specification.

The quality department may also take prime responsibility in setting up data and document storage systems for the collection, assembly and presentation of all documentation necessary to demonstrate the technical status of the construction as a whole at the time of completion. If this is not achieved satisfactorily, then the regulatory body will probably be unable to issue the necessary authorisation to operate the platform required by the government of the waters in whose national jurisdiction the installation is placed. This latter task is traditionally the role of a certification engineer but, as we will discuss in the next chapter, the QA group is increasingly becoming involved in this role.

But there are reasons other than statutory for which which full documentation of the structure is required. Modifications to the structure during its life, changes for instance in the process requirements which may arise and indeed damage incurred from fire, weathering or collision with supply vessels and ships, for example, must be regarded as possibilities and there will also be a necessary monitoring of the structual condition during its life with maintenance of the plant and structure. For this to be achieved competently, engineers, brought in to perform repairs, maintenance or modifications arising for any of these reasons, must be able to locate the dossier of documents, quality control reports, material test certificates etc., and more important, of any concessions given in the construction, so that the work they do can proceed on a sound basis of technical knowledge.

USER'S QA GROUP: DUTIES AND ACTIONS

From the technical aspects discussed in the above sections and the general nature of the contractual relationships and responsibilities indicated, a number of specific duties which fall upon the QA group for a project and some actions which should then be undertaken by them can be stated:

1. The placing of orders for materials, equipment and services can go badly wrong and, therefore, the QA group should make an early contribution to the procurement of materials, equipment and services by liaisons with the purchasing department and ensure that all purchase orders have the following:

(i) A statement defining the inspection requirements of items to be supplied. This must indicate clearly who will perform the various inspections, the standards or quality levels to be met and at what stages of production it will occur. Reference to ruling project specifications should be included.

(ii) The name of the prime contact in the user's QA department with whom the supplier should liaise (never let the supplier say he couldn't contact you).

(iii) A statement to prevent uncontrolled subcontracting. The user must be informed and his approval sought before any supplier concludes a sub-order. Regular surveillance should catch this but the user should take the initiative if in any doubt.

(iv) A clear indication of any inspecting agency or certifying authority who must witness work and approve materials and results at various stages of production. The mechanism for calling in such a body must be clearly indicated.

2. The making of assessments of potential suppliers as early as possible again in liaison with the purchasing department and especially for long lead items and major fabricators. This should take place preferably before, or at the latest, during the period of specification writing (a parallel activity in project groups with which the QA group should maintain an involvement). If this is not done orders may be placed with unsuitable suppliers (for commercial or other reasons) and the QA department may eventually find itself asked to retrieve quality situations by conducting assessments on suppliers who already hold an order.

The appraisal of a potential supplier of equipment, fabrications or services very often commences by the user sending out an appraisal questionnaire to all potential bidders. The questionnaire ought to be prepared by the user's QA group and should cover items such as:

Range of products or services supplied

Company QA status (eg what quality system standards they are "approved" to)

Details of the quality management team

Arrangements for work planning

Procedural instructions for work operations

Document control

Calibre of production and inspection staff

Design or drawing office activity

Possession of standards and codes

Use and calibration of instruments and test equipment

Internal auditing

Approvals held

Production and inspection methods

Control of defective items

Corrective action procedures

Production, preservation and handling of products

Training of staff

The initial results of this will enable the user's QA group to decide whether the company should be visited for an audit with the aim of including them on a provisional list of suppliers who may be invited to tender for work.

The objective will be to determine ultimately the suitability or otherwise of the company, in terms of its technical capability, its physical capacity for handling the proposed work and its managerial and contractual status, especially with regard to the assurance of quality in its products or services to be supplied. In this process, the subsequent visiting audit by the QA engineers in the user QA department is all important and the first step in doing this is to get a copy of the contractors quality manual for appraisal.

A preliminary review of this should be made, with regard to any recognised quality system, such as is defined in a number of national standards—BS5750, NS5801, Z299 etc.—and this must be thorough and total for the large fabricator. But the user must be judicious in making this appraisal. As we have seen, the standards are general and all-embracing in their stated lists of system requirements. Depending on the size and nature of the contractor and of the work intended, all the clauses may not be relevant. Service inspection companies, for instance, do not require arrangements for control of manufacturing operations, but their documentation, testing procedures and calibration of instruments are of paramount importance. Similarly, the users should not expect a small producer of one-off specialist "avant garde" equipment to have a full quality system manual if the company consists of two enterprising PhDs with an innovative product working at the production bench themselves. But accountability is still required and the appraisal should seek documented evidence of testing procedures and calibration of any instruments or measuring equipment used.

A response to the contractor is appropriate after review of a manual or any kind of QA submission so that the contractor has an opportunity to improve his system or prepare for the particular features of the project. For instance, the prospective supplier can be warned that he will have to establish particular arrangements for complying with the overall project scheme of controls, such as the submission of quality plans and inspection plans, an organisation for site co-ordination to cover aspects such as site queries, chart liaisons and interfacing, manning and planning of work etc.

Audits by the user on the manufacturing works or construction yards of prospective suppliers may therefore follow more or less rigorously the listed system requirements of established national standards but would typically cover:

Quality system operating

Quality manager—calibre and status in relation to parallel production staff

Organisation

Duties and responsibilities

Record keeping

Control of manufacture—procedures and inspections applied

Control of procurement and sub-contracting

Control of non-conformances

Concessions and corrective actions

Final clearance and certification

A similar sequence of events would be used by principal suppliers in placing sub-contract orders with material suppliers or service companies and, in all cases, the prospective bidders would be required to make a written documented commitment to respond formally on all such system details in the specified project format, as required during the time of their contracted involvement. Naturally, this assessment of a potential supplier's capability can be expensive and time consuming so the advantage of access to a previous audit of a supplier can be considerable. This is, of course, the value of certification schemes for suppliers and the, slightly different, QUASCO scheme (see chapter 2).

Technical assessments of vendors often fail to be fully effective or realistic because of the specialist technical knowledge and experience often required. Whilst QA departments are usually good at vetting inspection and NDT agencies they must not be afraid to enlist specialist technical assistance from other departments or from external sources. For instance, sub-contractors for welded fabrication must be assessed by a good welding engineer with managerial experience in a manufacturing company. Ideally, he should be a permanent member of the user's QA team for the entire project.

Audits of suppliers must be fully reported and aim to make a full and accurate appraisal of a potential supplier's true technical and managerial capabilities (or lack of them) as a solid basis for procurement selection and later contractual control purposes. Audits only have any value if properly reported and recorded and the conclusions should be stated clearly and without ambiguity including any data that was available. If any doubts exist a re-assessment can be made at a later date. No attempt should be made to use some kind of artificial scoring system by which the potential suppliers can be compared. Such methods are not suitable for major engineering projects and can give very misleading results.

Vendor assessment reports should also indicate clearly any non-runners and give clear guidance to procurement in the selection of favoured suppliers and most importantly they should include estimates of the probable and possible extra costs and delays that should be expected if certain suppliers are used. For instance, by incurring extra surveillance work by the user. Reports should also propose the level and extent of inspection or surveillance that will be necessary for the user/owner to use during fabrication and on completion before acceptance. These estimates can then be used by the QA department in drawing up its forward plans for site surveillance work and the associated manning levels required.

3. The preparation of documented procedures. As indicated above, one of the principal methods of control which the QA department must use is in the drawing up of procedural documents and specifications regarding particular technical aspects of work and their subsequent inspection and acceptance. A typical list of some of these controlling documents is given below and serves to exemplify the range and scope of an active quality group acting on behalf of an owner/operator.

1. QA vendor assessment guidance procedure for QA/QC systems.
2. QA inspection surveillance procedure for mechanical equipment erection and installation.
3. QA inspection surveillance procedure for fabricated pipe-work.
4. QA inspection surveillance procedure for node fabrication.
5. QA inspection surveillance procedure for non-destructive testing.
6. Instructions for installation and pre-commissioning of instrumentation.
7. Instructions for installation and pre-commissioning of electrical apparatus.
8. Jacket fabrication, documentation system.
9. Specification for a diving recompression facility.
10. Instruction to contractors for information recording format.
11. Pipeline procedure for welding and repair of pipelines to be used on the laying barge.
12. Specifications for field welding, coating and non-destructive testing of off-shore pipeline systems.
13. QA instructions for operation of computer terminal.
14. Structural fabrication specification.
15. Repair procedures for underwater welding.
16. Tender documents for jacket and module installation and hook up work.
17. Construction specification for fabricated piping and insulation.
18. Steel for topsides structures.

Clearly from the content of these titles the documents are intended for primary use and guidance by different parties involved in the project. Thus 1–5 are primarily for the use of the user's own appointed QA staff at the departmental base and at each construction site and they define the scope of staff activities and how they are to perform the surveillance duties required.

Site fabricators will have been required to perform their own quality activities and the QA surveillance defined will be fundamentally a verification that the supplier is correctly fulfilling his obligations and performing the required tests, measurements or inspections correctly. This may well involve reference to a previously approved detailed inspection procedure that the supplier has submitted and is committed to use, and it will also cover the correct and complete recording of QC results to the format required.

Documents 6–14 are essentially intended as instructional documents for the sub-contractor's guidance and effectively spell out to him what he must do and how he is to do it. They and the project specifications will form, together with other ruling codes, standards and procedures, the contractual basis for the project. They usually therefore have their origin within one of the technical groups of the project. Large oil companies may well have the in-house technical expertise to generate such documents as project specifications or instructions but it is not uncommon to find that specialist companies or consulting engineers may be contracted by a project group to create these essential ruling documents.

Documents such as 11 and 15 define detailed technical operations and sequences and will often have been originated by the engineers of a specialist sub-contractor who has been selected and appointed to perform the work. They will therefore have been drawn up following detailed verbal discussions with the user, indicating what is required in broad terms and also with due regard to any overall ruling project specifications, eg the structural fabrication specification, 14. They would then have been submitted to the user for approval, a process which may involve the opinion of other external bodies such as engineering consultants or certifying authorities and then reissued as project ruling documents back to the contractor. Procedures for special heat treatments, non-destructive testing, application of special coatings such as metal sponged aluminium or the installation and commissioning of electronic control or communications equipment would fall into this category.

Document 16 is an example of a document given to all prospective contractors for a specific item of work, indicating the scope of work required, referencing the ruling documents, probably some of those above, and telling him how his tender is to be presented. This latter aspect is important to the user as it can ensure that certain essential aspects of a tender are presented in precisely the same way by all tenderers, thus making the bid appraisal job easier and more realistic in comparative terms.

Document 18, specifying structural steels, would be one of the first documents generated by the user as the long term ordering of steel for the construction as a whole depends upon it. Normally based on a national steel specification the document will include a range of extra requirements which the user or his representative, often a firm of engineering consultants, deem to be prudent in the light of past experience or the particular characteristics in service of the intended platform.

All specifications and procedural documents and instructions must be issued by the project manager or director in accordance with a formalised system of control and approval and to an established standard format. The heading sheet should indicate as a minimum; the document title giving a meaningful, concise and unambiguous indication of what the document is all about, a unique identification number (which will probably consist of mixed letters and numbers and accord with an established company or project system for the numbering of documents—this would probably include the project identification but if not this should also appear on the title sheet), the date of initial issue, the copy number, the revision status of the document (for which Revision 0 is often used for its initial issue) and the page number with total number of pages in the document. Subsequent revisions with the pages involved and dates of issue are usually detailed on a revisions page somewhere inside the document.

Also indicated on the heading sheet should be signatures and dates for those individuals involved in its preparation and approval. Typically these will be the originator or the person by whom the document was proposed; this may sometimes be endorsed by a supervisor or, if appropriate, the head of department or project technical group by which he is effectively approving the document's contents for issue. This may be the user's QA manager if it is a QA surveillance document; if not, then the QA manager would sign it to indicate that all quality-related aspects have been covered correctly and that conflicts with other project documents or certifying authority requirements have been resolved or eliminated, and finally the project manager or director would sign. For some documents it will also be necessary to have the formal endorsement of the certifying authority involved, especially when they have been appointed to act also as a technical authority by their government agency or government.

The precise role of "overseeing" authorities can be very confusing and it is surprising how often many project and QA staff are unaware of the true contractual position and therefore authority of the "third party" involved. It may vary in different parts of the world with the national governments involved and whether they have government departments who wish to exercise an authoritative responsibility. Overseeing authorities may be involved as "inspection" authorities, as "technical" authorities or as "certifying" authorities and they may be contracted directly by governments, or move usually by the user/operators and sometimes by main contractors. In this case, though their involvement is most likely to be in the "inspection" role. All documents issued for a project should also carry certain items of standard information:

An "intended for use by" statement should appear either on the title page or in an introductory leading section or scope statement, usually on the first page.

The purpose, aims and objectives of the document should also appear in an introductory or "scope" section at the beginning of the document.

To place the document in context and indicate its contractual or mandatory weight, there should be an early reference to other related standard documents, codes or procedures which have a bearing on the application or interpretation of the document, and a clear statement as to which documents are overriding in the event of any dispute or conflict.

In association with these above statements there should be an explanation of terms used, in particular regarding the parties involved, such as the "Company", the "Client", the "Supplier", the "Works" etc., and also sometimes the locations and items of structure or plant involved.

Only when all possibility of ambiguity in terms and interpretations has been eliminated can we hope for the documents to be used readily and effectively. Even so, every project has its difficulties and disputes and so a mechanism for the referral of problems and provision for the competent resolution of them by the user must be built into the project management system as a whole. In the establishment and monitoring of this provision, the quality department has an important role to play.

4. The monitoring of contractors' quality control performance including sub-contractors they may use is the principal project activity of the user QA group and the site surveillance teams should be established as previouly suggested. This will involve teams of staff at each construction site having representative skills and expertise in welding engineering, visual inspection, non-destructive examination, materials control and documentation as appropriate. In practice, these teams may be led by individuals transferred from a head office QA group as the emphasis of the work changes. Manning levels should initially relate to an original assessment by the QA manager and should then be subject to adjustment during the project life to reflect the technical content of current work and the peformance of the contractor.

Every site team should be provided with clear instructional documents indicating the level of surveillance to be performed and giving details of the inspections to be applied, the ruling project specifications and how the results are to be recorded and reported to the project management. A typical duty for a site team is to stringently monitor the material quality identification and, when required, its verification, either in the supplier's works but, more usually, at the reception yards where the material will be stored in readiness for construction work.

In performing these surveillance duties the user's QA team must be careful not to take away the fabricator's responsibilities by over-inspection. Clearly, the fabricator has a contractual duty to control the quality of work produced by means of his own planned inspections and tests. Surveillance by the owner to verify that this is done is usual and appropriate but if quality control breaks down or technical problems relating to workmanship arise at the construction site it is tempting for the user to get involved in detailed inspection and testing alongside the manufacturer's inspection team. If this happens there is a danger that the manufacturer may step back and let the user get on with it and effectively replace his own quality control. This takes from him the incentive to make the necessary improvements and incidentally relieves him of some contractual responsibilities. Over-inspection like this can kill the supplier's initiative for action and must be resisted by the user; extra efforts should rather be focused on the QC procedures and personnel rather than on the actual hardware.

5. The maintenance of quality records. The site surveillance team manager must maintain documented records of all site QA activities and rigorously monitor progress in resolving any technical problems or non-compliances which are indicated by the appropriate technical staff or project groups within the user's organisation.

In this connection, the QA staff must refrain from making technical decisions of acceptance–rework–reject significance because that is primarily the responsibility of the project technical groups. Exceptions to this may occur when the QA department has individuals with specific knowledge and experience but in general the QA department should not act as an internal technical consultancy. It is not in the direct line of technical responsibility.

As a futher word of caution, if a serious technical problem arises mid-way through a construction programme, then the response from the user should be to emphasise and rigorously apply the quality control measurements and methods specific to the item or problem rather than to attack the contractor's quality assurance system as a whole. For instance, a spate of welding problems half way through a project should not be the signal for the user to demand a new QA system from the fabricator. Probing critically into matters like the records of internal audits and purchasing, calibration and stores and demanding new manuals to be written will only help the next client. It is too late for that to have positive effect on the current job and corrective action must be directed to the quality control staff, procedures and inspection records on which the final stamp of approval for acceptance will be won. Remember, it is quality control rather than quality assurance which will raise quality in situations controlled by levels of workmanship.

This collection and control of all documented records of inspection work has been mentioned above relating to the build up of a dossier of information for final presentation to the Certifying Authority or other regulatory authority. In fact this will invariably occur on a progressive basis as the project proceeds and the quality assurance group will need to control this liaison so that the necessary approvals are formalised—as will be discussed more fully in chapter 5. There is scope in this activity for computer storage of all records and reports so that rapid feedback of information and details of repairs etc. is possible. Increasingly, because of the amount of documentation involved, this computerisation is becoming more prevalent. This, of course leads to concerns about the QA of the computer software itself.

6. The maintenance of communciation on quality matters between the participants. The acquisition of data and information should place the QA group in a strong position to promote effective communications between all groups and ensure that there is an effective feedback of information and in the checking of decisions made. This implies, of course, that the staffing of the QA group is correct, both in terms of numbers of people and their skills, to enable this to happen.

However, at the earliest stages of a project there are several aspects which can be effectively promoted and monitored by the QA department in its liaison with the design group. As we have seen above, the geological considerations of the sea-bed and the local environment where the platform is to be placed, give rise to the size limit for the platform and the weight restriction. This is one of the more important factors which can be monitored and controlled by a QA group.

As we have seen from chapter 3, the quality assurance system must ensure that all companies and design departments involved in the project for major items such as the jacket or modules are fully aware of the weight limits to which they have to operate. The quality department must be especially diligent in ensuring that this restriction is applied not only in the structural steel work but also in the design of pipe work and vessels which are contained within them. The means by which this can be achieved is in the form of procedural operations and guides which can be conveyed to the several design departments, but it is essential that this consideration is formalised and built into the methods of routine working and liaison early in the life of the project.

However, when construction is underway, the activity changes somewhat in that the principal involvement for the quality department is in the application of quality *control* monitoring in the fabrication yards where the weight accumulation is carefully logged and this must be introduced as an item in the final dossier of records and technical history of the project as a whole. It should be remembered in this respect that the concept of weight limitation persists throughout the design life of the structure and not merely up to the point of installation and commissioning. It is rare, indeed, for any offshore installation not to have some modifications or additions made to its structure during its lifetime and a detailed record of the weight distribution would be essential in planning any such changes.

Material selection is a subject often dominated in the offshore situation by the low temperature characteristics of the environment or the particular corrosion resistance requirements which apply. The QA group can make a major contribution here by ensuring that the necessary range of skills is collected from within the user organisation, or indeed with external specialist advice to assist the relevant design groups. Structural designers and corrosion engineers are rarely also experts in welding technology and non-destructive testing.

Similarly, with regard to design and the methods of construction to be used, the involvement of quality assurance is similar to its role in material selection. Major decisions have to be made at the early design stage and the quality management should ensure that aspects of design such as welding procedures to be used, methods of construction, the position of joints and their configurations are made with due regard to the necessity for performing welding procedure qualification tests and to the methods of inspection or non-destructive examination which will have to be applied. In both these instances design differences may materially affect the potential effectiveness of the welding operation or the methods of inspecting it. There have been many welded joints designed into constructions in the past which have proved to be devoid of all access necessary to perform the welding operation or else cannot be inspected by any of the recognised NDT methods.

The construction phase activity then becomes more a matter of quality control or, perhaps, more correctly, surveillance of the supplier's quality control activities.

THE QA GROUP'S INVOLVEMENT IN PROJECT CONTROL

When construction work or service company activity is underway, there are a number of ways in which the user's QA department can play a part in assisting the project management teams as well as carrying out its main role of monitoring the "quality" effort.

It is usual for contracted work at all levels to be monitored in its progress by weekly project meetings and these should always be attended by representatives of the QA group or site team. The user with his QA manager will call in his contractor who will also have his QA or QC people in attendance and the contractor will monitor the service company in the same way.

The QA group must ensure that items of concern are included in the agendas of these meetings and the results of all the QA surveillance and monitoring can find expression at these meetings, at best in simply recording the achievements and their level of formal contractual approval and, at worst, in identifying the technical difficulties encountered or reject or repair situations so that matters can be effectively resolved by the full project authority. At the same time reapproval of working progress and volume of activity on a regular basis enables manning levels in the QA site surveillance teams to be adjusted in good time.

A further important control that is applied relates to the use of inspection and quality plans. Contractually, the user is in the position of specifying and defining a number of technical requirements, documented as we have seen in a range of documents with unique identification and numbered clauses.

A quality plan requires the contractor to tabulate against each requirement to which he has to respond by some action or quality management arrangement the number of the document or other means of responding, the identity of the person responsible for carrying it out and the means of logging or recording the result as a permanent record. The quality plan is literally a map of the requirements and responses relating to one documented specification and it carries provision for the various contracted parties and the certifying authorities to indicate what level of surveillance they wish to apply to each item (they may insist on work being held until they have cleared it, operations are to be witnessed by them or subject to patrol inspection only). Figure 4.2 is a typical example of part of a quality plan. As can be seen, this is a significant document and is the "core" document controlling the quality activity on a project.

An inspection plan is a similar document but is based not on a specification from the user but a technical procedure of work operations submitted by the contractor (perhaps for approval by the user). It therefore lists the operations involved in applying the procedure generally in order of sequence and provides the opportunity, again, for all inspection or test stages to be stated with provision for holding, witnessing or inspection by the client or certifying/inspection authority. Again, a typical example is given in Figure 4.3.

CONCLUDING REMARKS

It was stated at the beginning of this chapter that the construction and installation phases are the most difficult to control. The message on quality assurance must therefore be to put as much emphasis as possible on a

	ACTIVITY CODE	REVISION		O	A	B	C	NAME	FORM NO.
	P = PATROL INSPECTION M = MANDATORY INSPECTION H = HOLD W = WITNESS R = REVIEW BY INTERNAL AUDIT A = CERTIFYING AUTHORITY APPROVAL REQ. C = CLIENT APPROVAL REQD.	APPROVALS DATE & SIGNATURE							
			CLIENT						
			CERT/A.						Sheet 1 of 1
CLIENT:	PROJECT:	DOC. REF: SUB-SEA ANTI-FOULING I.D. MARKERS.							

STAGE	QUALITY REQUIREMENTS	ACTIVITY	CLIENT SPEC. REF.	RAMCO PROCEDURE REF:	Proforma No.			ACTIVITY CODE				SIGN OFF
							CL			CL	C/A	
1.	Q.A./Q.C. System to be established and maintained.	Appointed Manager for Site quality. Quality plans and Inspection plans submitted.	3.0 3.1					M				
2.	Perform material supplied proving tests.	Tests on adhesives and materials etc., supplied at goods inwards stage. Frequency, one per batch or delivery unit. Acceptance levels to be manufacturers claimed valves:	3.1.2					M				
3.	Inspect all panels and fixing materials for character, dimensions and quality as specified.	Subject to inspection by quality control supervisor.	3.1.3.					M				
4.	Supply five year performance and two year prior storage guarantees for fixing items.	Included in documentation package defined by procedure.	5.0					M				
5.	Supply documentation as specified.	Includes strap and adhesive test data in addition to 4.0 above.	6.2					M				

Fig. 4.2 Typical layout of quality plan

	ACTIVITY CODE	REVISION		O	A	B	C	NAME	FORM NO.
	P = PATROL INSPECTION M = MANDATORY INSPECTION H = HOLD W = WITNESS R = REVIEW BY INTERNAL AUDIT A = CERTIFYING AUTHORITY APPROVAL REQ. C = CLIENT APPROVAL REQD.	APPROVALS DATE & SIGNATURE							
			CLIENT						
			CERT/A.						Sheet 2 of 4
CLIENT:	PROJECT:	DOC. REF: SURFACE PREPARATION & COATING.							

STAGE	ACTIVITY SEQUENCE	INSPECTION ACTIVITY	CLIENT SPEC. REF.	INSPECTION REPORTING.	Proforma No.		ACTIVITY CODE				SIGN OFF
									CL.	C/A	
7.	Inspect blasted Surface.	Visual to SA3 and Profile Depth.	"	Q.C. Supervisor			M				
8.	Vacuum Clean	Visual	"	Q.C. Supervisor			M				
9.	Sling and Move	Visual	"	Q.C. Supervisor			P				
10.	Metal Spray Bay	Cleanliness Inspection	"	Q.C. Supervisor			M				
11.	Plant and Metal Wire	Visual and Batch Nos.	"	Q.C. Supervisor			M				
12.	Metal Spray equipment function.	Test piece for spray pattern.	"	Q.C. Supervisor			P				
13.	Metal Spraying	Visual	"	Q.C. Supervisor			P				
14.	Surface Quality & Thickness Check.	Instrument and Visual	"	Q.C. Supervisor			M				
15.	Work move to Painting Bay	Visual	"	Q.C. Supervisor			P				
16.	Pre-Sealer Inspection	Cleanliness Inspection	"	Q.C. Supervisor			M				
17.	Plant and Sealer Check	Visual and Batch Nos.	"	Q.C. Supervisor			M				
18.	Environment	Instrument Check	"	Q.C. Supervisor			M				

Fig. 4.3 Typical layout of inspection plan

systematic way of monitoring and controlling the quality and make maximum use of standardised forms of documentation.

The actual techniques of control (ie the manufacturing procedures, the inspection and test procedures) are matters to be decided by the technical specialists in the appropriate disciplines. In a typical offshore project there will be, as has been stated, an immense variety of types of engineering and manufacturing activity, each with its own traditions, rules and jargon. The role of the quality assurance staff is to control these activities in a managerial sense to ensure the desired level of quality is being achieved and to demonstrate (particularly to regulatory bodies) that it has been. A major part of the quality assurance role is therefore maintaining control and information flows across interfaces.

It is paradoxical that the construction phase is the phase of a project which is most easily visualised as being amenable to the quality assurance discipline and yet is the one which causes most of the observable problems. Partly this is because of the sheer scale and diversity of the activities involved but partly it is because of the traditional inspection oriented nature of the industry. Much of what has passed for QA in the construction phase of offshore projects has really been traditional inspection. Much of that has to remain but it is necessary, if only for economy of effort, to move as far as possible towards ensuring that both users and suppliers have adequate quality systems in place to prevent mistakes and failures rather than rely on an army of inspectors.

REFERENCES

4.1 *The Effect of Welding Procedure Qualification Requirement on Production Costs.* NEDO Books, London 1985.

4.2 Bentley, K. P. *Quality Assurance in the Fabrication of Offshore Structures Welding in Energy-Related Projects.* Pergamon Press (Canada) 1984.

4.3 Grant, I. M. and Rogerson, J. H. *The importance of contractual requirements in determining quality costs in the fabrication industry.* W I Conference "Fitness for purpose Validation of Welded Constructions", London 1981.

CHAPTER 5

CERTIFICATION, SAFETY AND RELIABILITY

J. Weston

INTRODUCTION

Any industry in which failures can have very great consequences in terms of safety is naturally subject to "special" controls imposed by governments. The offshore industry, in common with the nuclear power industry, the mining industry and the aircraft industry is one of these. In all of these industries, governments seek to protect the public, as far as is practicable, by establishing rules which operators must comply with before they are allowed to operate.

In the offshore industry this activity is called "certification" and is, quite obviously, closely related to quality assurance. We might even define quality assurance, in a rather narrow way, as the means by which an operator demonstrates that his installation meets Certification requirements. However, because of the history of the industry, Certification has developed independently of quality assurance. This seems, with hindsight, absurd and the message now is that the two activities should be integrated as far as is possible as they both share similar objectives achieved through similar routes.

To start any discussion on the subject of Certification, Safety and Reliability it is important to define the meanings of the terms as they are understood by the offshore oil and gas industry. "Certification" has evolved to mean "those series of activities and records required to satisfy a country's regulatory authorities that the safety and reliability of construction and operation of an offshore installation in their national waters are adequate and comply with defined standards". Expressed differently, the word Certification can therefore be taken to be synonymous with the words Safety and Reliability and in this chapter Certification is discussed in this context.

It is also important to understand how the processes of Certification operate. In order to satisfy any bureaucratic authority responsible for the administration and control of Certification that the required standards have been adhered to, the owner or operator must present evidence of the quality of design and construction of the installation. Thus Certification can become very closely aligned with quality assurance (QA). Preliminary design, detailed design and specification, fabrication and construction, vendor appraisal, installation, hook-up and commissioning, operation/ maintenance and re-certification are all subject to scrutiny and approval by both QA Engineers and the bureaucratic surveyors responsible for Certification. Certification is a continuous process from initial design through daily operation/maintenance of the completed installation to statutory re-certification which is required to maintain the installation's approval to operate—its "Certificate of Fitness".

To achieve the level of answerability which is required in the Certification process it is critical to maintain comprehensive documentary records at all stages of an offshore project. Documentation control is not normally regarded as an engineering function but, for the Certification Engineer, the need to be able to assess documents for their legislative compliance is matched by the requirement to be able to manage the often voluminous project records to ensure their accessibility.

Overall Certification is a multi-discipline, project-wide function which requires both dedicated and general support during all stages of an installation's life. It is an essential and vital part of any offshore hydrocarbons production project but perhaps surprisingly it is not as precisely defined as one would expect. In certain areas detailed specifications are laid down but in others the requirements are general. For example, in the UK, the Offshore Installations (Operational Safety, Health and Welfare) Regulations 1976 SI 1019 define the required contents of the medical cabinet down to the number of safety pins but leave the Offshore Installation Manager the scope to appoint "...a sufficient number of competent persons to be responsible for..." the main structure, essential electrics, drilling operations and so on.

THE BACKGROUND TO CERTIFICATION

It must be understood that much of the detail of Certification schemes around the world tends to be reactionary, that is the specific regulations have evolved over the years in response to particular accidents or major disasters. This approach is not, of course, unique to the offshore industry. Before an examination of some of the dominant aspects of Certification can be made it is therefore useful to trace the history of Certification schemes in general in the maritime business.

The first form of Certification developed in the 18th century when a common interest in "ensuring" standards of safety and reliability was identified between shipowners, shipbuilders and insurance underwriters.

Each had a different bias. For example, the owner wished to minimise operating costs (including his insurance premiums) and wanted to be sure of the standard of vessel that he was buying from the shipbuilder. The shipbuilder wished to enhance his reputation by marketing ships of defined standards (note here the analogy with the different "interests" in quality assurance as discussed in chapter 1). This tripartite group met in a coffee house in London owned by an Edward Lloyd and through time his house became the centre of shipping information. Gradually lists of ships were compiled giving guns carried, cargoes, crews and owner's name, tonnage etc. Grading the ships A to E for the standards of the hull and 1 to 5 for the standards of the equipment enabled the underwriters to increase or decrease premiums as required. To define this grading or "Classification" Lloyd's Register of Shipping was founded in London in 1760. The first "Classification Rules" were published in 1835 for wooden ships and the rules for iron ships were produced in 1855. The Lloyd's term "A1" became an international industrial symbol of quality and safety.

From these pragmatic beginnings, far from the realms of government legislation, the business of "Classification" has developed throughout the world as a means of ensuring, predominantly for commercial advantage, the application of acceptable standards for the construction and operation of ships. The Classification Societies not only publish the Classification Rules but they also maintain, often worldwide, networks of surveyors to oversee their observance. Apart from Lloyd's Register of Shipping, the major Classification Societies are:

Bureau Veritas (Paris)

American Bureau of Shipping (New York)

Germanischer Lloyd (Berlin)

Det Norske Veritas (Oslo)

Certification for shipping has grown from its 18th century origins and now includes, over and above Classification, National Legislation and International Regulations. These developments are significant because offshore Certification of the oil and gas industry has grown out of Shipping Certification and to understand its philosophy, structure and limitations an appreciation of its origins is important.

Disasters at sea have always received widespread publicity, and every major incident or series of casualties stimulates public and governmental pressure for improved safety and more stringent regulation. For example, the loss of 548 British ships and 3000 lives in 1882 resulted in the introduction of the first UK legislation to preserve the "freeboard" of ships by restricting cargo loading to the famous Plimsoll line. At the same time concern was being expressed over the high frequency of shipping losses through collisions. The net result was the first International Maritime Conference which was held in Washington in 1889. The Conference concentrated first on establishing rules for the prevention of collisions, the "rules of the road". The Conference agenda was comprehensive and included a wide variety of safety at sea issues but it was decided that no international rules could be made at that time in respect of seaworthiness, that a uniform system of load lines was not ripe for consideration, and that the establishment of a permanent international Maritime Commission was not then expedient.

A further major disaster, the tragic loss of the *Titanic* initiated the next step which was the calling of an international Conference which, although delayed by the first world war, was eventually held in 1929. Since this disaster was the result of a collision with an iceberg which caused the vessel to sink through extensive flooding with the loss of many lives because of the inadequate provision of lifeboats, the 1929 Conference was concerned primarily with the sub-division of ships and the provision of life saving equipment. The conclusions of the conference (and subsequent conferences) led to the formation of the International Maritime Organisation (IMO).

The IMO is now one of the specialist agencies operating within the framework of the United Nations Organisation. It is the principal body responsible for international Certification rules and regulations, both for shipping and for offshore hydrocarbon activities, the most important of which are:

1966 Load Line Convention (which established the Plimsoll line internationally).

Safety of Life at Sea Convention (SOLAS).

Convention on the International Regulations for Preventing Collisions at Sea.

International Convention for the Prevention of Pollution from Ships (MARPOL 1973).

Code for the Construction and Equipment of Mobile Offshore Drilling Units (MODU Code).

For a ship, the application of these international Certification rules and codes is dependent upon its country of registration. If the shipowner chooses, for example, to register his vessel in Liberia, then only those conventions which have been formally adopted by Liberia apply to the vessel. Similarly, a British registered ship only has to comply with the UK ratified conventions. This flexibility to choose, in effect, the Certification scheme under which the shipowner is prepared to operate, has given rise to the expression "flag of convenience", which aptly describes the limits of international regulation.

For an offshore installation operating in the "controlled waters" of a particular country, the system is inevitably less flexible. By the simple requirement for offshore installations to register their operations in "controlled waters", the country develops the means to treat the installation in an analogous way to a ship registered in that country. British registered offshore installations are treated somewhat like British ships, Canadian offshore installations somewhat like Canadian ships and so on.

The extension of shipping Certification methods to offshore installations owes much to the similarity of the circumstances surrounding their introduction. For example, just as shipping legislation and international regulation followed notable maritime disasters, the UK Certification scheme for offshore installations was established following a major accident in which 13 lives were lost when a jack-up exploration platform—the Sea Gem—overturned in the North Sea in the early 1960s.

THE UK LEGISLATIVE FRAMEWORK

Faced with the requirement for a rapid response to public and industry pressure and with no legislative precedents arising from onshore exploration, the approach adopted by the British Government was to examine its existing resources and introduce what is known as an "enabling" Act in 1971. This Act (The Mineral Workings (Offshore Installations) Act 1971) empowered the Secretary of State to issue subordinate regulations detailing the legislative regime that offshore operators were to follow. Borrowing from shipping legislation the Secretary of State introduced the concept of an appointed independent "Certifying Authority". In his search for an appropriate body (or bodies) to carry out this independent supervision, the Secretary of State turned to the Ship Classification Societies (with their pre-existing sets of rules and trained surveyors) to carry out the new function.

The prime responsibility of the Certifying Authority is defined in Offshore Installations (Construction and Survey) Regulations SI 289, 1974 as "...to be satisfied that the measures taken by the owner to comply with his statutory obligations (under the regulations) are adequate and effective and, when so satisfied, to issue a Certificate of Fitness".

This general, all-embracing requirement encompasses the more detailed specific requirements that were made within the provisions of the enabling Mineral Workings Act. In outline, the structure of current UK offshore legislation is illustrated in Figure 5.1. The highest level is an Act of Parliament, which is normally general in content, but which "enables" the Minister responsible (ie the Government Department) to issue more detailed regulations. The most important Acts applicable to UK offshore installations are the Mineral Workings (Offshore Installations) Act 1971 and the Petroleum and Submarine Pipelines Act 1975. There are also four further Acts which impinge on offshore activities.

STRUCTURE APPLICABLE TO ALL UK LAW		DETAILS OF OFFSHORE LEGISLATION		NOTES
STATUTORY AND QUASI LEGISLATION	HIERARCHY	DEGREE OF LEGISLATION	SPECIFIC HIERARCHY EXAMPLE	
STATUTORY	ACTS OF PARLIAMENT	CIRCA 25 ACTS	MINERAL WORKINGS (OFFSHORE INSTALLATIONS) ACT 1971	
	STATUTORY INSTRUMENTS	CIRCA 75 S.I.'s NOT INCLUDING SAFETY ZONES	14 STATUTORY INSTRUMENTS: 1972/702, 1972/703, 1972/1542 1973/1842 1974/289, 1974/388 1975/1289 1976/1019, 1976/1542 1977/486 1978/611 1980/1759 1983/1258 1984/419	
NON - STATUTORY	GOVERNMENTAL - GUIDANCE NOTES - C.S.O.N.'s - SAFETY LETTERS/ALERTS - D.S.M.'s - M.S.N.'s	- OVER 65 RELEVANT PUBLICATIONS - 49 C.S.O.N.'s - 110+ LETTERS - 175+ D.S.M.'s - 30+ (RELEVANT) M.S.N.'s	18* GOVERNMENTAL PUBLICATIONS VARIOUS C.S.O.N.'s VARIOUS SAFETY LETTERS, NOTICES AND ALERTS NO D.S.M.'s ** APPROX 10 M.S.N.'s	* NB THIS INCLUDES DOCUMENTS WHICH HAVE BEEN PREPARED SPECIFICALLY TO EXPLAIN A SET OF REGULATIONS AND OTHER DOCUMENTS NOT PREPARED SPECIFICALLY BUT COVER COMMON SUBJECT MATERIAL ** DIVING SAFETY MEMORANDA, THOUGH PREPARED BY THE DEPARTMENT OF ENERGY, ULTIMATELY DERIVED FROM THE HEALTH & SAFETY AT WORK ACT 1974 AND ARE THEREFORE NOT UNDER THIS PARTICULAR HIERARCHY
	NON GOVERNMENTAL - INTERNATIONAL LAW & CONVENTIONS - ENGINEERING STANDARDS - CODES OF PRACTICE - SAFETY & GENERAL	- VARIOUS - VARIOUS - VARIOUS - OVER 30 DOCUMENTS	- VARIOUS - VARIOUS } 24 DOCUMENTS	

Fig. 5.1 UK Legislation

The detailed subordinate regulations, that is to say the regulations issued under an enabling Act, are Statutory Instruments (SIs). There have been a number of SIs issued over the years to regulate various aspects of offshore operations. The most important to the UK Certification process are:

SI 289	(Construction and Survey) Regulations—which provide the basis of the Certification scheme: the appointment of the Certifying Authorities and the definition of the Certificate of Fitness procedures and processes.
SI 1019	(Operational Safety, Health and Welfare) Regulations—which provide for the application of the relevant regulations of the Health and Safety at Work Act offshore.
SI 486 & SI 611	(Life Saving Appliances) and (Fire-Fighting Equipment) Regulations—adopted substantially from shipping legislation, these provide for the review and examination of safety equipment on offshore installations and are totally separate from the Certification of Fitness. The Certificates of Examination for both Life Saving Appliances and Fire-Fighting Equipment are issued by the Marine Survey Officer of the Department of Transport.
SI 1398	(Prevention of Oil Pollution) Regulations—this merchant shipping based legislation arising from an International Convention is applicable in part to offshore installations. It requires the examination and Certification of certain waste water treatment plant.

In support of the legislation the appropriate UK Government department also issues a variety of quasi-legal instructions in the form of Continental Shelf Operations Notices, Merchant Shipping Notices, and Guidance Notes (generally issued following an incident or accident). These are designed to further inform the operator of the requirements of the UK Government and, whilst not having the full force of law behind them, there is an implied duty for the operator to conform or alternatively justify any non-conformance. It is this guidance to which inspectors and surveyors from the Certifying Authority and other regulatory bodies will generally refer when undertaking their assessments.

It should be noted that not all Government issued regulations fall within the responsibility of the Certifying Authority to ensure operator compliance. In the case of Life Saving Appliances and Fire-Fighting Equipment, for example, Certification is carried out by Department of Transport Surveyors. It is nevertheless part of the function of a Certification Engineer working for an operator to be fully conversant with the varying requirements of the different legal instruments in order to ensure correct compliance.

Because of this range of requirements to be met, in addition to meeting the requirements of the Certifying Authority, the operator needs to make applications seeking approval from other regulatory bodies at various stages throughout the development of an installation. Applications must be made in good time to allow the authority sufficient opportunity to assess the implications and request further information or clarification of the application. There are well over 20 "Authorisations" required for a typical offshore installation and the assessment of the application for Authorisation will take either a few hours or a few months. Reference 5.1, which is a guide to UK offshore legislation, itself runs to more than 250 pages. It is therefore vital that the Certification Engineer appointed by the operator understands the bureaucratic processes so that he can assist in the smooth passage of the review and approval of applications.

Figure 5.2 gives details of a few of the Authorisations required within the UK scheme.

THE CERTIFICATION REQUIREMENTS IN OTHER COUNTRIES

Other countries operate very similar regulatory regimes, with greater or less emphasis on the external supervision of compliance and, obviously, with slightly different statutory requirements. The Norwegian system is worth examining here since it is one which places emphasis on self-regulation and quality assurance.

The Norwegian regulations are somewhat different from the UK regulations although, taken as a whole, achieve similar results. The Norwegians are, however, advanced in their use of risk assessment techniques as a support for the legislation, this being due in part to the type and nature of the major accidents in the Norwegian sector of the North Sea in the past ten years—another example of how regulatory requirements tend to be developed in response to specific incidents.

The "Guidance for Safety Evaluation of Platform Conceptual Design" is produced by the Norwegian Petroleum Directorate and directs the operator on how to attain Certification objectives, by providing specific guidelines on:

Document requirements

Methods for performing analysis

Basic safety standards

As with most Norwegian offshore regulations it is the responsibility of the operator to show that he has carried out an internal safety evaluation and has, if required, altered his design to minimise risk.

The legislation itself covers many of the same areas of installation con-

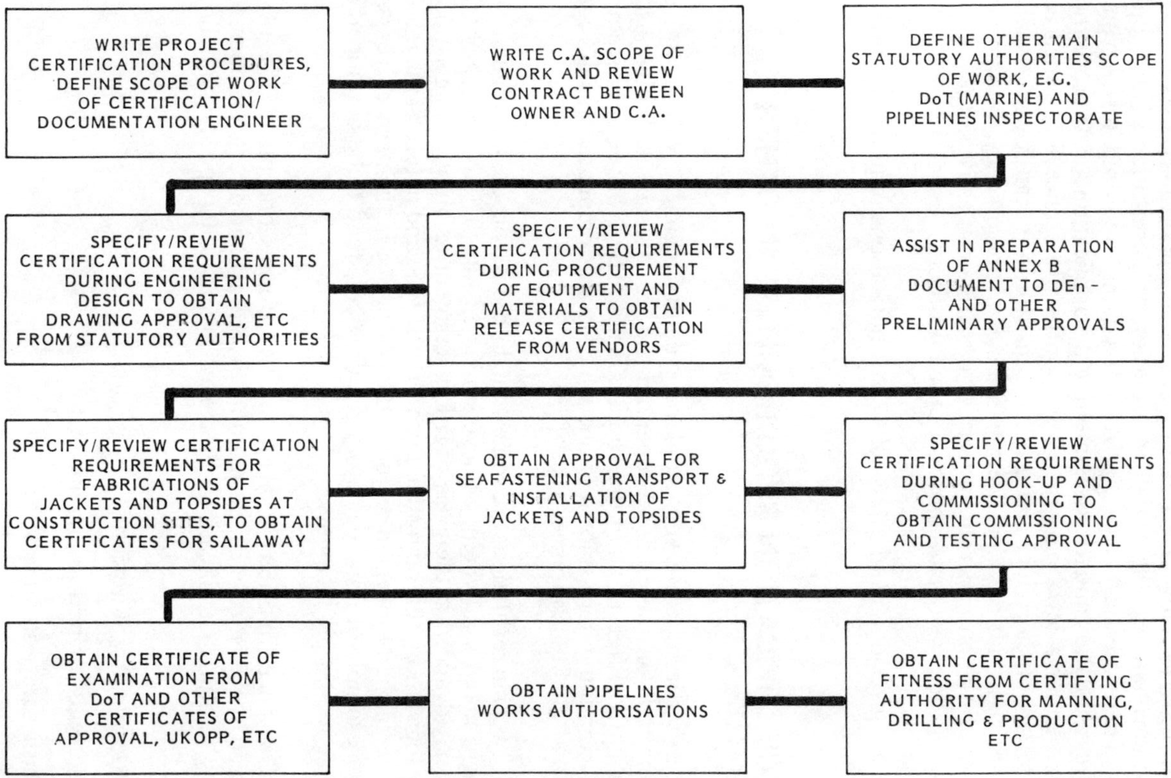

Fig. 5.2 Certification milestones

struction and operation as the UK legislation but there are differences in its structure and application.

As in UK legislation, the primary Norwegian legislative instrument is an Act, (see Fig. 5.4) which is general in nature and "enables" the introduction of more detailed regulations. It is in these detailed regulations that the Norwegian legislative system differs: firstly, there are "Regulations Stipulated by Royal Decree", which establish the basic requirements of, and operation of, the Norwegian regulatory system; secondly, there are "Regulations Stipulated Pursuant to Royal Decree", which detail the specific requirements of the Norwegian government, eg Structural Design of Loadbearing Structures, Cranes on Production Installations. Finally, there are a number of Guidance Notes covering certain specific design and operation activities, eg safety evaluations, inspection of primary and secondary structures.

However, the major difference between UK and Norwegian legislation lies in its method of application. The "Regulations Concerning the Licensee's Internal Control ..." places responsibility on the operator to control and undertake his activities in such a way as to ensure the safety, etc of his installation and operations. This is commonly termed "self-certification".

Although at first glance, this system may appear to allow much more freedom than the UK system, the operator is still subject to a number of regulatory processes (or consents) designed to assess how he performs his activities. The "Consent to Operate an Installation", for example, is only valid for six months (against five years for the UK Certificate of Fitness). A good example of the operation of the Norwegian system was provided by the case of the drilling semi-submersible "Henry Goodrich", which was refused Consent to Operate, when newly arrived from its Japanese fabrication yard, due to inadequate documentary records.

The term Self-Certification requires the operator to show by the use of good internal quality assurance techniques that he can reach the required standards of design, construction, operation and maintenance without using a Certifying Authority. This places *full* responsibility on the operator and the control systems he utilises. The legislative system effectively audits the management ability of the operator to ensure that the declared design is actually achieved, installed and safely operated.

The contrast in philosophy between this method and the UK approach can be summarised in terms of the objectives of the scheme:

The objective of the UK system of assessment by independent Certifying Authorities is to ensure adequate standards of safety offshore by compliance with minimum or mutually agreed standards of design, construction and operation.

The objective of the Norwegian system is to provide a situation where, by minimal interim regulation, the operator is required to define the highest possible standards in his activities in order to ensure that he obtains and maintains his Consent to Operate.

Obviously the Department of Energy (UK) and The Norwegian Petroleum Directorate each claim that their differing approaches are the best but it is probable that the net result of either approach is little different if properly followed. Operators under the Norwegian system will seek to comply with whatever minimum acceptable standards that they can identify, be they legislative, agreed industry standards or guidance notes published by the independent assessment bodies (such as Det Norske Veritas). Alternatively, the operator under the Norwegian system may choose to employ an independent assessment body to provide guidance and continuous assessment throughout the project: a "certifying authority" in all but official title.

The prime problem with the Norwegian scheme is that the operator in theory cannot guarantee satisfaction of Norwegian Petroleum Directorate (NPD) requirements until their inspectors arrive in response to his request for Consent to Operate; in practice he will undoubtedly maintain informal contact with the NPD throughout the project development. This will take many forms, for example; spot checks and audits (to see that procedures are being adhered to); visits by surveyors to witness testing of equipment; design and calculation reviews.

A close liaison must be therefore developed in this self-certification regime so that no holdups occur. As in any scheme the planning procedures written at the start of the project must be good enough to allow for all conditions that may arise during the development of the project.

SEQUENCE OF APPROVAL

UK	*Norway*
UK Legislation	Norwegian Legislation
Certifying Authorities	Internal Control
Certificate of Fitness	Self-Certification
Certificate of Examination	Consent to Operate

Through their different routes the UK and Norwegian Certification Schemes arrive at their common objective—the safest possible operating environment offshore. The role of an independent appraisal body is a fundamental tool in any Certification and safety scheme.

In conclusion, it is important to recognise that a Certification scheme is never static: its rules and regulations are constantly modified as operating experience, technical knowledge and the increasingly sophisticated analytical and appraisal techniques are translated into regulations. Innovative designs will result in new controlling procedures, just as disasters and accidents will force the refinement of guidance notes and bring new areas of activity within the scope of the law.

KEY STEPS IN CERTIFICATION

Good planning is the key to success in many activities. This is certainly applicable to Certification activities. Of equal importance is good records control, since Certification processes rely to great extent on correct, verified documentation. We can see, therefore, the similarity in approach to the control of project management and design (chapter 3)—in fact, "Certification" is one section of the whole project management operation.

The importance of documentation control is underlined by an appreciation of the quantities of documents involved in the design, construction and hook up of an offshore platform, prior to the commencement of operations.

During the design phase the volume of paper is primarily due to the generation of specifications, procedures, design drawings, purchase orders, and sub sea reports. Most of these may appear to have little to do with final Certification but it must be remembered that the Certificate of Fitness is only the final certificate at the top of a huge pyramid of documentation. All documents required to initiate and control a project are also essential to the final Certification of the installation.

For example, fabrication in a major project can involve 5–10 different sites each having different ways of managing and documenting their activities. This leads to different material numbering systems. Each site utilises weld procedures, welder's qualifications, NDT operator's qualifications, site procedures, site drawings, transmittals, weight reports, dimensional reports, material records, copies of the specification, design drawings, purchase orders, vendor documents and many more different forms of records. The procedures written at the design stage for handling this documentation often fail to cope with the sheer variety and volume of documentation involved on the various sites—ie the interface problem as discussed in earlier chapters.

Vendor documentation is defined as the records produced by a vendor to demonstrate that his equipment conforms both to the regulations and to the requirements of the purchase order/specifications. Little attention has often been paid to stipulating to Vendors precisely what documentation is, and more importantly since Vendor documentation can be vast, what is not, required for certification purposes. For the Certification Engineer to comb through large volumes of Vendor documentation, searching for the precise details required by the Certifying Authority, is time-consuming and inefficient. Close liaison between QA and Certification is also essential if the process is not to be duplicated between the two functions.

Hook up and commissioning have historically been offshore activities but the economics of offshore operations are forcing as much commissioning as possible to be done on site onshore to reduce the offshore phase. Hook up, ie the joining together of systems and the modules built on different sites and brought together offshore, leads to testing of the completed systems. Such testing together with the use of additional materials and equipment offshore results in further documentation some of which is critical for the final Certification processes.

Figure 5.3 indicates the typical generation profile and numbers of documents involved in the project phase of an offshore installation. It can easily be seen from the total numbers of documents produced that good procedures need to be applied to control records and retrieve Certification information right from project commencement.

The physical bulk and weight of Certification data have led operators to extend their use of computers and microfilm storage/retrieval devices. A clear and sound information management policy encompassing Certification appraisal at source and long-term data storage—whether by microfilm, hard copy or computer disk—is now an essential part of offshore Certification. There is little point in developing a procedure which covers the generation of data during construction and installation but fails to cover the basic information retrieval requirements during operation. Certification legislation imposes a statutory obligation to be able to demonstrate *at all times* the standards of design and construction.

This introduction of computers for information management purposes has helped to control Certification information but even with these advances the Certification Engineer will always need to see the bureaucratic stamp and signature on a certificate for verification.

Working within the operator's or main contractor's project team, the prime responsibilities of the Certification Engineer are to ease communications with the various regulatory bodies, to contribute to the appreciation and understanding of the complex legislative and engineering concepts involved in offshore developments and to ensure that the obligatory bureaucratic processes and safeguards are satisfied in a proper and expeditious manner. The role is an important one, since failure to produce the correct Certification has a knock on effect. A "hold" imposed by late production of key documentation can lead to late arrival of equipment on site, consequential non-payment of outstanding accounts, delays with sub-vendors and so on.

The breadth of the Certification Engineer's responsiblities require that, in the performances of his duties, he undertakes a variety of roles, including

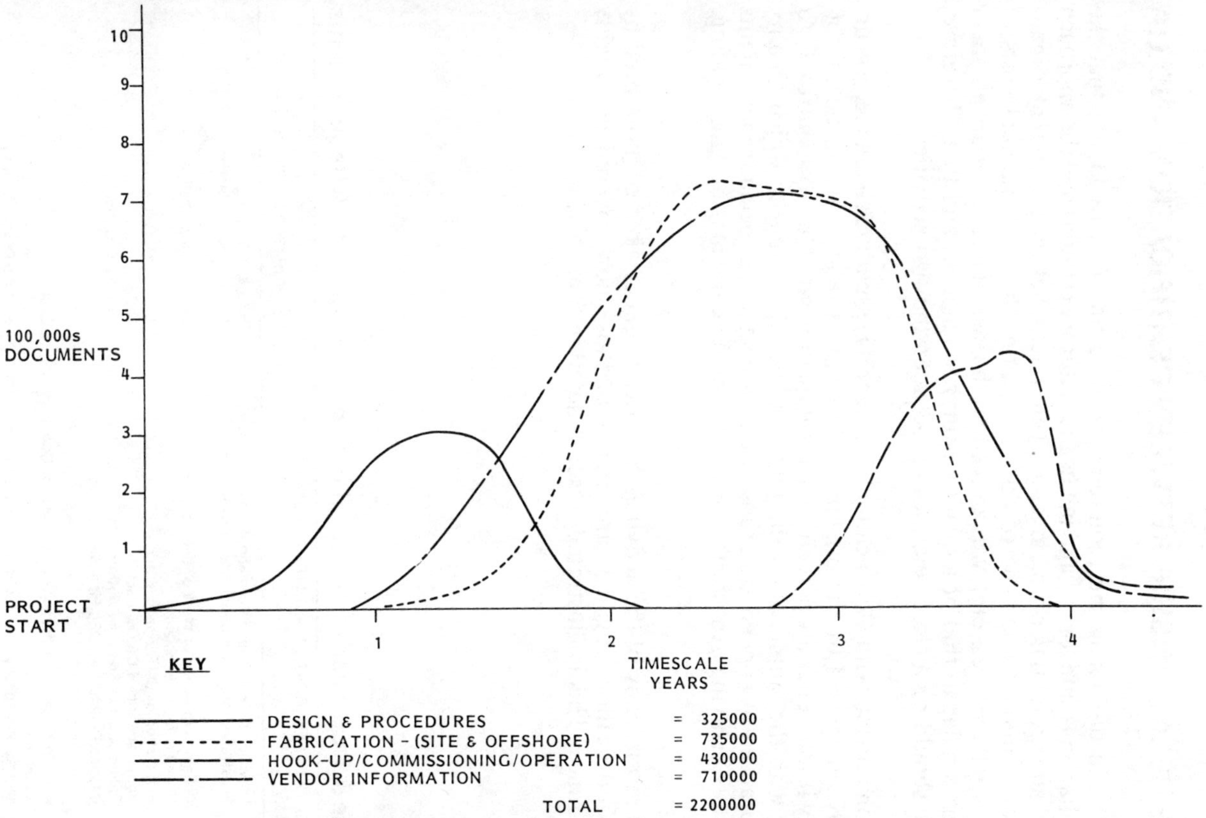

Fig. 5.3 Document estimate for a typical offshore project

Quality Assurance, Engineering Support, Quality Control, Documentation Control and Information Management.

THE RELATIONSHIP BETWEEN CERTIFICATION AND QA

So far, in discussing the requirements of certification (evidence that the installation is safe for use and meets the statutory criteria) and the approach used and potential problem areas (interfaces between different organisations, amount and type of documentation, the consequential costs of delays in the process etc), we can see that the task of a certification engineer is very similar to that of a quality assurance engineer and that, therefore, there should be a close relationship between the two activities.

In comparison with Certification Engineering, Quality Assurance, as currently recognised by the offshore industry, is a relative newcomer. Originally it was generally perceived in combination with Quality Control (QA/QC) with the emphasis on the QC elements. The areas to which Quality Assurance was applied were those areas subject to the specific requirements of the Certifying Authority, eg review of fundamental design, welding, materials.

As a result many of the activities historically undertaken or monitored by the Certification Engineer are similar to those now covered by Quality Assurance. This is illustrated in some detail in table 5.1.

Table 5.1. Comparison of Certification Activities and QA Requirements

	CERTIFICATION ACTIVITIES/REQUIREMENTS	*QA REQUIREMENTS (BASED ON BS 5750 PART 1)*
1.	Provision of legislation information and interpretation	Sect 4.10 Training
2.	Preparation of Project Certification procedures	Sect 4.1 Quality System 4.5 Work Instructions
3.	Definition of Scope of Work of Certification/ Documentation Engineer	Sect 4.2 Organisation
4.	Preparation of CA Scope of Work	Sect 4.2 Organisation
5.	Administration of CA Contract	N/A
6.	Define of Other Statutory Authority Involvement, eg DoT (Marine Directorate), DEn Pipelines Inspectorate	Sect 4.2 Organisation
DESIGN		
7.	Input to project specifications and codes for Certification requirements	Sect 4.4 Planning 4.5 Work Instructions
8.	Identify and specify design documentation subject to CA Review	Sect 4.8 Design Control
9.	Review documentation for certification compliance prior to CA submission	Sect 4.8 Design Control
10.	Management of the CA design submissions ensuring that all returned comments are logged, distributed and actioned	Sect 4.9 Documentation and Change Control

Table 5.1 (contd)

CERTIFICATION ACTIVITIES/REQUIREMENTS		*QA REQUIREMENTS (BASED ON BS 5750 PART 1)*
PROCUREMENT		
11.	Specify certification documentation and CA involvement requirements for purchased equipment and materials	Sect 4.5 Work Instructions
12.	Audit/review vendor certification	Sect 4.6 Records 4.11 Control of Purchased Material and Services
13.	Management of non-conformance procedure, e.g. spec waivers	Sect 4.7 Corrective Action 4.16 Control of non-conforming material
14.	Establishment of recept, inspection and certification ident schemes, eg 'Clean' warehouse	Sect 4.11 Control of Purchased Material and Services 4.16 Control of non-conforming material
15.	Consolidation and submission of the Material/Equipment Certification to CA for review	Sect 4.6 Records
FABRICATION		
16.	Review/approval of material traceability procedures	Sect 4.13 Purchaser supplied material 4.16 Control of non-conforming material
17.	Review/approval of welding/manufacturing/rectification procedures	Sect 4.12 Manufacturing Control
18.	Review/approval of Inspection/Testing Procedures	Sect 4.14 Completed item inspection and test 4.17 Indication of inspection status
19.	Review of Welders/NDT Operator's Qualifications	Sect 4.19 Training
20.	Review of Inspection/Test equipment calibration records	Sect 4.10 Control of inspection, measuring and test equipment
21.	Co-ordination of CA survey activities to obtain release certificates prior to sailaway	N/A
22.	Management of non-conformance procedures	Sect 4.7 Corrective Action 4.16 Control of non-conforming material
23.	Co-ordination and Review of Fabrication Records	Sect 4.6 Records
INSTALLATION AND COMMISSIONING (OFFSHORE)		
24.	As items 16–23 offshore construction activities	As above
25.	Co-ordination of CA survey activities during Commissioning	Sect 4.14 Completed item inspection and test
26.	Obtain Certificates of Fitness for appropriate stage of completion, eg Occupancy, Drilling, Production	N/A
27.	Assembly/Review of Installation/Commissioning Records	Sect 4.6 Records
OPERATION AND MAINTENANCE		
28.	Co-ordinating CA Annual and Major Surveys	N/A
29.	Co-ordinating submissions to CA for changes to installations, maintenance and inspection reports	Sect 4.6 Records
STATUTORY OPERATIONS MANUAL		
30.	Co-ordinating preparation and submission of original manual to CA for approval (*Note:* Certificate of Fitness Requirement)	Sect 4.6 Records
31.	Updating and re-submitting Statutory Operational Manual to include revisions to installation	Sect 4.9 Documentation and change control
LIFE-SAVING APPLIANCES *FIRE-FIGHTING EQUIPMENT*		
32.	Co-ordinating submissions of design documentation to DoT (Marine Directorate)	Sect 4.8 Design Control
33.	Co-ordination of offshore survey visits for issue of Certificate of Examination	Sect 4.14 Completed item inspection and test
34.	Co-ordination of any interim and the bi-annual re-surveys required to revise and renew the Certificate of Examination	Sect 4.9 Documentation and change control
PIPELINES		
35.	Co-ordinate preparation/submission of Pipeline Works Authorisation request	Sect 4.8 Design Control
36.	Co-ordinate attendance of Pipelines Inspectorate for hydrotesting	Sect 4.14 Completed item inspection and test
37.	Collect and collate (as appropriate) the Pipeline installation records and post installation survey	Sect 4.16 Records

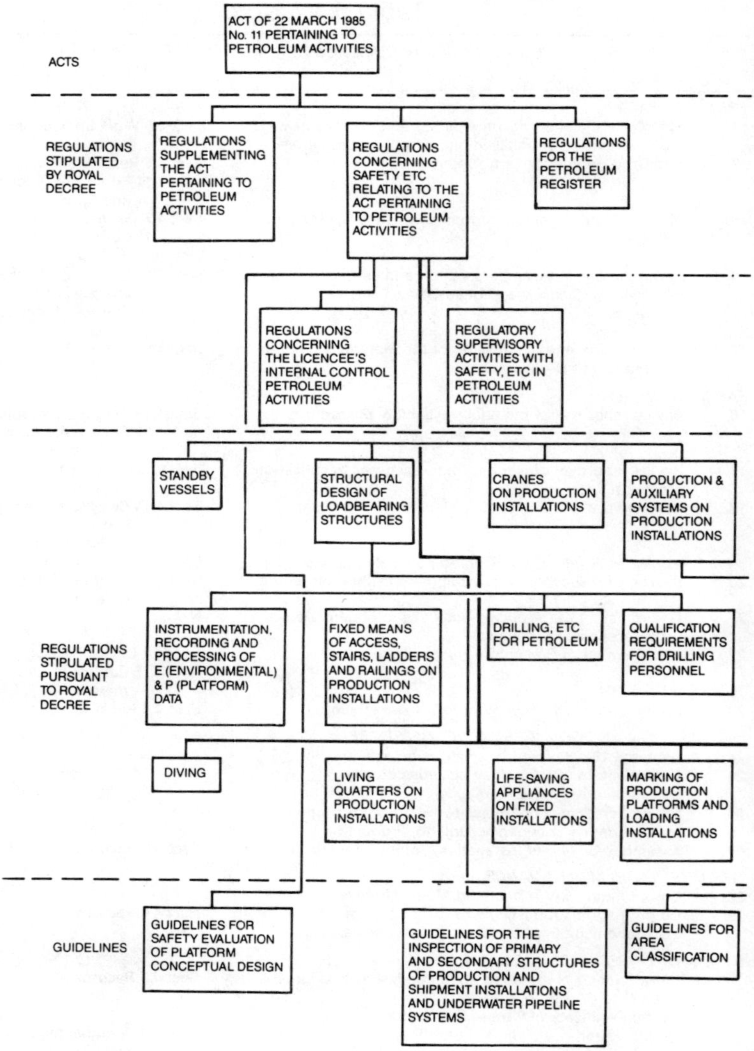

Fig. 5.4 Norwegian

The prime difference between Certification and QA which still exists, and will continue to do so, concerns each discipline's ultimate purpose. QA is primarily concerned with ensuring that the methods of working are correct to produce the "right" finished product. Over and above this "quality" element Certification is also concerned to ensure that verification is provided on paper with the correct approved signatures, in order to satisfy the regulatory authorities. Therefore, certification is a specific objective of a project (unless certification is achieved the installation cannot legally oper-

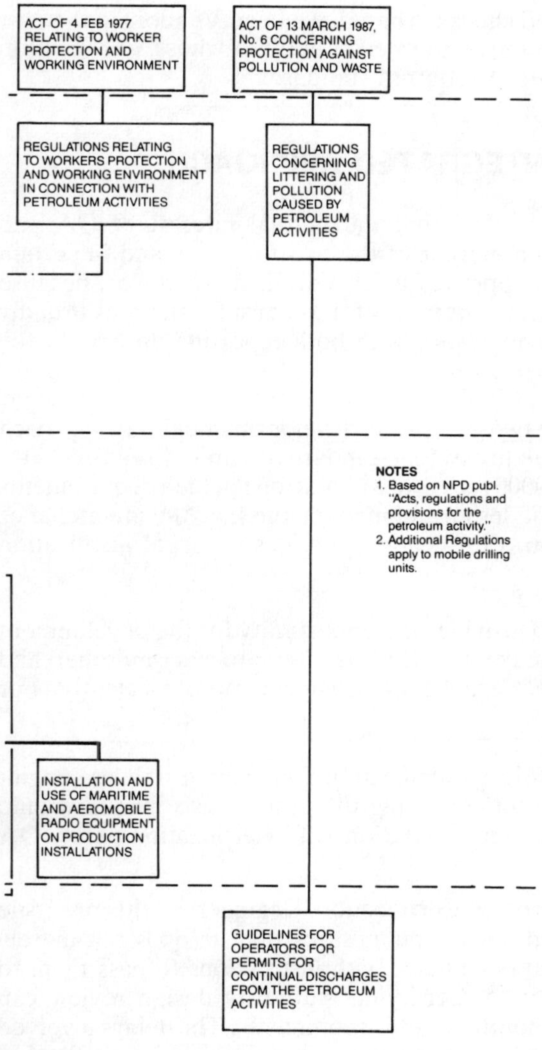

offshore legislation

ate) whereas quality assurance is a means of ensuring that a project is correctly completed.

Certification and QA are thus two somewhat separate project activities with differing objectives but sharing a mutual interest in the way many other project group activities are performed. For example, when Certification and QA personnel conduct Vendor appraisals both look for good procedures and control in-house which will give confidence to the engineers when they start to conduct their detailed checks. There have been Vendors in the past who had excellent QA procedures and personnel, but whose Certification documentation fell well short of required standards.

THE VALUE OF AN INTEGRATED APPROACH

It has been suggested in the UK that the mutual interest of QA and Certification should lead to a merger of the two functions and in certain projects in recent years this approach has been taken. However, because of their different objectives, it is more useful to consider them as broadly *parallel* activities of equal importance with both reporting directly to the Project Manager or equivalent.

With this latter approach the two groups can develop an integrated approach to the project. The various quality system standards (chapter 2) eg BS5750—Quality Systems or the ISO 9000 series, are very strong in their requirements for verification at all stages of design and manufacture and the maintenance of associated documentation. Certification provides the legal justification for such requirements.

An integrated approach also provides the opportunity for the development of project procedures to make the Certification process smoother and to use the expertise of the Certification Engineer more efficiently. For example:

(1) The inclusion of the Project Certification Engineer in the pre-engineering team responsible for preparing the design basis manual would allow for the proper representation of Certification and QA requirements.

(2) The inclusion of the Project Certification Engineer in the pre-issue review process of design documents ensures that he does not merely receive the issued design documents with instructions to pass them to the Certifying Authority. A Certifying Authority design review can take weeks, sometimes months to return comments. The delays involved can frequently be avoided through contact by the Project Certification Engineer during the internal review stage, particularly when the Authority requires the inclusion of some design changes.

(3) It is common in main design contractors for the field QA inspectors responsible for inspecting goods at vendors works to sign off, but not approve, the Certification documentation accompanying it. As a result the Project Certification Engineer ends up with a massive team of Certification engineers, QA engineers, expeditors and documentation engineers trying to chase up and approve vendor Certification prior to release of equipment. This situation could easily be avoided if the Certification documentation was regarded as an integral part of the purchase order for the goods themselves and was checked *and approved*

prior to shipment, either by the QA inspectors or visiting Certification Engineers. A similar approach towards the receipt and inspection of goods at the fabrication sites would also improve efficiency.

(4) A major project may utilise anything between 5-10 fabrication sites plus a larger number of sub-contractors. To ensure accessibility and compatibility of the fabrication record books, the project Certification Engineer will generally prepare a procedure detailing how these records should be assembled and ideally providing a recommended index format.

To improve this situation, the procedure should be written into the contract (and sub-contracts) and it should be a part of the site QA function to ensure that the fabricator (and his sub-contractors) comply with it. The site and offshore Certification Engineers will then have an established standard for all documentation which will result in greater traceability and improved retrieval of information at a later date.

(5) Inclusion of the Project Certification Engineer into the Hook-Up and Commissioning planning team will allow him to identify the requirements for the Certifying Authority (or any other Regulatory Body) to attend testing programmes. It will also enable him to ensure that procedures for compiling the Hook-Up and Commissioning Records have been executed in accordance with the planned programme.

Towards the end of a project it is all too common to find that the Certification Department needs a manning level approaching that of the design engineering department, made up of Certification Engineers, QA Engineers, field and in-house documentation expeditors, all trying to attain the required standard of Certification documentation. Their task is frequently made more difficult because the "required standard" was not clearly specified in the early stages of the project.

(6) The Certification Engineer can, if involved at the right stage in the project develop the required Information and Records Management policies.

SAFETY AND RELIABILITY

The Safety aspects of Certification for an offshore project can be divided into two functions for each phase of the project:

1. Safety in the engineering design of the installation.

2. Safety of personnel.

The "Assurance of Quality" of the Safety and Reliability of offshore installations is generally provided through the Certification process with the reviews and assessments being undertaken by the Certifying Authority and other regulatory bodies. Where there is a Loss Prevention discipline group within the design team, the QA department may however become involved in the formulation and monitoring of safety reviews, Hazard and Operability (HAZOP) studies and Hazard Analysis (HAZAN) studies.

Safety has its origins in design and strictly does not finish with completion of construction. Safety, like Certification, crosses all project activities, from design of handrails for essential machinery to hazardous area classification. For those engaged in design, construction or production the broad requirements for safety continue as the project advances.

In the UK offshore industry particularly, the safety requirements of Certification have been given special emphasis in the last ten years.

In September 1978 the UK Health & Safety Commission agreed to the establishment of an ad hoc working party: "To consider and advise on proposals for regulations and, if appropriate, Guidance Notes for the protection of people at work from hazards to health and safety from their occupation in the construction industry involved in the construction, demolition or dismantling of an offshore installation as defined in Article 4 of the Health and Safety at Work Act 1974."

Part 1 of the Health and Safety at Work Act 1974 (HSWA) now applies to the construction of installations offshore in the United Kingdom Continental Shelf and in territorial waters (although it should be noted that these are Guidance Notes and therefore non-statutory regulations). The Notes were derived partly from onshore construction experience and partly from the NWOIA/Shipbuilding and Ship Repairing Regulations 1960 (again the legislation has used the precedent of other marine law). The purpose of the Guidance Notes is to recommend standards for compliance with the appropriate sections (2–7) of the Health and Safety at Work Act.

Responsibility for satisfying the standards fall on the person having the obligation to comply with the provisions of the Act.

The UK Offshore Installations (Life Saving Appliances) Regulations SI 486 1977 relate to the requirements for, and maintenance of, life-saving appliances for offshore personnel and the equipment is subject to reviews by the UK Department of Transport who issue the relevant certificates.

Life saving covers all the obvious areas: survival craft, liferafts, lifebuoys, lifejackets, means of escape, etc. Each of these areas is appraised and approval or type approval is essential before the item is fitted offshore. Offshore there are subsequent tests on, for example, davits and winches to check loadings, rate of descent, followed by retesting on a regular basis to maintain the valid certificates.

The UK Offshore Installations (Fire Fighting Equipment) Regulations SI

611 1978 are another set of rules devoted to maintaining safety offshore. This particular UK statutory instrument covers such areas as detection systems, fire fighting equipment, gas monitoring and detection equipment, and sprinkler systems. Recertification and retesting of equipment are again compulsory.

Safety requirements also extend to personnel training. The hazardous environment offshore and the increasing sophistication of survival and fire fighting equipment and techniques make personnel training essential. For the UK sector a one week course covers life saving, survival techniques and fire fighting. It is, in general, subject to renewal procedures after a set period of time when personnel are required to go on refresher courses to up-date their knowledge and retest their skills.

Within the scope of safety comes Risk Analysis, Hazard and Operability (HAZOP) and Hazard Analysis (HAZAN). These are areas that are becoming very specialised on a worldwide basis due, primarily, to the rapid advances in technology and the introduction of new materials with unknown properties. Potential risk situations are thereby created for which past experience is of little or no consequence.

The studies in these specialised areas that are being conducted not only help in the creation of safer work environments, but the findings of such studies also highlight critical areas of concern. If a particular situation is rectified early enough the risk of accidents is reduced, and this in turn can lead to a reduction in the frequency of certain types of hazardous events.

Risk Analysis, HAZOPs and HAZANs are important areas of development in the context of offshore safety and Certification. The appraisal of a design by the Certifying Authority or the adoption of a design in a self-certification regime must involve an assessment of "risk". The definition of a risk is "exposure to mischance" whilst the definition of hazard is a "source of danger". When the two aspects are being considered together hazard analysis is deemed to be a restricted form of risk analysis.

Hazard analysis is often used to identify hazards in industrial processes, in particular through analysing failure modes, and is the most researched and formalised form of risk analysis. It involves a team of specialists who will identify the hazards in a particular process system and subsequently make recommendations to reduce to an acceptable factor the probability of danger. Rules and regulations lay down certain criteria, but do not identify or attempt to evaluate the problems involved in meeting the requirements. The objective of Hazard Analysis is to meet the safety objectives embodied in the legislative framework.

Risk Analysis attempts to highlight the interdependency of items within the overall scope of a design. Evaluation of the risks is conducted by a fault tree analysis or event analysis depending upon the risk being investigated.

Background statistics are used to help in certain areas in determining the value of the risk involved eg: adverse environmental conditions.

Risk Management is measured in terms of:

Accidents to personnel

Environmental impact

Property damage

Health problems

Cost via loss of business

In this, offshore risk analysis is no different from the risks of analysis carried out at, for example, nuclear power stations and any other processing plant. Where risk analysis offshore differs is in the use it makes of the pre-existing Certification categorisation of the component parts of the installation. For example, within the UK Certification system there exist "category ratings" for different types of equipment. These can be used to help pinpoint high risk areas:

Category "A"—Failure could be a danger to the platform or personnel, eg major structure of pressure vessels. Full design appraisal and a full survey by the Certifying Authority at the equipment manufacture is required.

Category "B"—Failure will not affect overall safety, but will lead to individual system failure. Design appraisal by the Certifying Authority is not required, but independent survey at the manafacturer's works plus witness tests, running tests and approval of certificates is required.

Categories "C" & "D"—No critical effect will be caused by failure—manufacturer's documentation and test certificates are sufficient without independent appraisal.

It can be seen from the list above that Category "A" items constitute the greatest risk and hence require the greatest amount of QA and Certification effort. All equipment must be installed on a UK offshore installation to the satisfaction of the Certifying Authority surveyor and built to a recognised international code or specification. Examples of the categories and the certification requirements in the UK are shown in Figure 5.5.

Again, we can see the similarity between "category ratings" for certification purposes and "criticality ratings" for QA purposes (chapter 2). The point of each of these ratings schemes is to identify, in a semi-quantitative way, the most critical parts of an installation which can them be subject to more detailed scrutiny (a hazard or risk analysis in this case).

UK Certification (Documentation and Inspection) Requirements For Equipment and Materials				
Category	*Equipment*	*Documentation Required for C.A. Design Appraisal*	*C.A. In-Works Inspection*	*Certification Documentation Required*
A	Pressure vacuum vessels, tanks, heat exchangers, pit traps or fired equipment. Storage tanks for hazardous or toxic fluids. (Small vessels, filters, drain traps, etc., of standard manufacture may be classed as category (iii).	Design and fabrication specifications, drawings, calculations, bill of materials, weld procedures, Q.C. proposals.	Yes	Data book comprising: Material certificates Weld examination and NDT reports Heat treatment reports Hydraulic or other test reports
B	Pressure relief valves	None	Yes	Valve certificates including material certificates, pressure and setting test reports
C	Life saving and safety equipment (e.g. lifeboats, liferafts, etc.)	None	No	Manufacturers certificates of compliance and certificates of type approval of recognised testing authority
D	All equipment and materials that do not fall into any of the above categories and are of a routine or domestic nature	None	No	Manufacturers certificate of compliance only

Fig. 5.5

Reliability Analysis does not strictly fall within Certification schemes although this may change shortly (at least in the UK system).

Reliablity Analysis deals with the probability of events—eg, events which may cause loss of production, damage or loss of life. Most studies of this nature have been carried out on UK offshore production platforms with the principal objective of defining production levels for use in financial forecasts although some limited work has been done on the basis of predicting structural failure using probabilistic techniques taken from the nuclear industry (ref 5.2). The reliability of the facilities is predicted against the design "ideal" to give an idea of what can actually be expected. The breakdown for the Analysis is by system, equipment within the system, the failure rates expected for each item and the restoration time then expected to prepare, repair and recommission the equipment. In the USA a data-base system is well established, but despite recommendations by governmentally sponsored working Committees, the equivalent in the UK and most other offshore areas does not yet exist. The overall maintenance of reliability data can provide an important starting point for both re-certification exercises (to maintain existing Certification) and for the initial Certification of offshore projects.

The American regulations are covered in APIRP14C for the Analysis Design, Installation and Testing of Basic Surface Safety Systems offshore. A detailed flow chart of the process is produced from which the operating details of equipment design and process requirements are listed. Checklists are then drawn up to evaluate items of equipment, both as "stand alone items" and as items interdependent on other pieces of equipment. A Failure and Inventory Reporting System (FIRS) was set up in 1980 with the prime function of information collection. The operators are required to report inventory and failure data which are then stored and collated. The information is made available to manufacturers and operators and highlights areas of safety concern where equipment has given problems. In the long term the collated information will eventually help both equipment designers and operators to build and maintain more reliable equipment.

This is another instance, in fact, where the QA approach (failure analysis leading to improved preventive measures) is used.

REFERENCES

5.1 Whitehead, H. UK Offshore Legislation Guide, published by Benn Technical Books.

5.2 Wong, W. K. and Rogerson, J. H. *A probabilistic estimate of the relative value of factors which control the failure by fracture of offshore structures*. 2nd International Conference on Offshore Welded Structures. The Welding Institute, London 1982.

ACKNOWLEDGEMENTS

The author acknowledges the considerable assistance given in preparing this chapter by the technical staff of Edward J Weston & Co. Ltd.

CHAPTER 6

QUALITY ASSURANCE IN OPERATIONS AND MAINTENANCE

J. H. Rogerson

INTRODUCTION

It is obvious that a complex, expensive and potentially hazardous installation such as an offshore oil and gas platform and its associated facilities needs to be maintained and operated in a proper and safe manner. Consequently, it seems equally obvious that if a formal quality system approach to design, manufacture and installation is necessary then a formal quality system approach to operation and maintenance is also necessary.

To date, however, the quality assurance emphasis in the industry (indeed in most industries) has been on design and manufacture and there has been very little concerted effort to apply the quality system approach beyond these stages. The quality system standards, as was stated in chapter 2, reflect this in that the only one which makes *specific* mention of operations, maintenance (and de-commissioning) is the "nuclear" standard BS 5882. Certainly, the stress on all other standards (including the new international standards, the ISO 9000 series) is on design and manufacture with a somewhat perfunctory reference to the provision of "services" and, in the case of ISO 9001, "servicing". The format and content of these standards naturally reflect the concerns and priorities of industry.

There may be a number of reasons for this apparent lack of concern about assuring the quality at the post-manufacturing stage. These might include:

1. Operators need to control their suppliers in a "quality" sense but do not perceive the need to control their own operations in the same way as they assume "normal" management and engineering controls will see to that.

2. The cost of lack of quality in operations and maintenance may not easily be distinguished from the cost of the inevitable plant modifications and repairs which arise from operating conditions and the

ageing of plant, ie there is no obvious driving force for a formal quality programme.

3. The activities themselves are so complex, interrelated and subject to so many external factors that it is thought impossible to apply worthwhile quality systems.

It is profitless to speculate on the real reasons except to say that similar reasons were doubtless employed 15-20 years ago when quality assurance as a formal discipline was first introduced to design and manufacture.

What we can say is that it is highly likely that formal quality systems will need to be imposed on these post-installation activities—if only as the result of pressure from regulatory authorities—and that we should analyse how it should be done and how (if at all) it differs from the approach currently used for the assurance of design, manufacture and installation.

THE TYPES OF ACTIVITY INVOLVED

We can divide the operational activities down into three basic types (which overlap to some extent).

These are:

1. The maintenance of the structure and facilities so that they are safe to use (ie the maintenance of the certificate of fitness to satisfy the relevant regulatory authority).

2. The maintenance of the structure and facilities so that they operate in an economic and efficient manner (ie the "normal" plant maintenance activities).

3. The operation of the facilities as process plant under normal (or emergency) conditions.

Note, we will not consider de-commissioning as an activity at this stage.

These three types of activity, although of course interrelated, have different primary aims and, therefore, the aim of a quality system will be different in each case. This could entail three different quality systems but this will depend on the management structures of the operator and how these activities are controlled.

For example, how much is controlled offshore on the platform and how much is controlled and directed from an onshore base? The control, on a day to day basis, of the operations will be offshore based but much of the other two types of activity can be (and sometimes is) controlled from onshore.

TABLE 6.1 Primary Aims of a Quality System

Activity	*Aim*
1. Maintenance of Certification	To demonstrate that a planned programme of inspections and tests has been satisfactorily carried out to satisfy a regulatory authority.
2. Maintenance of the facilities	To plan, control and record preventive maintenance, repairs and modifications needed to keep the facility operating efficiently and economically.
3. Operations	To provide and implement guidelines, procedures and standards to control the safe operation of the structure in normal and emergency conditions.

In a general sense it is more appropriate to state the aims of quality assurance of the three activities and then see how management systems should be directed towards these aims. Table 6.1 itemises the major *aims* of quality assurance in each of these three activities (note the overlap inevitable in this). As we go down table 6.1 we see that the aim of the quality system changes from an emphasis on providing documentary evidence to an emphasis on providing the active management of the activity. It is, of course, conceptually much easier to restrict "quality assurance" to the provision of paper work.

THE RELEVANCE OF THE QUALITY SYSTEM STANDARDS

The aims and the relative advantages and disadvantages of the various types of quality system standard have been already extensively discussed in chapter 2 but mostly from the rather traditional point of view of their application to design, manufacture and installation.

In chapter 2 we developed the idea of a matrix of features and activities to define, in effect, what a formal quality system aims to do and we then discussed the various standards in relation to this matrix concept. The five features of a quality management system—defined responsibilities, defined procedures, defined follow up action, adequate records and adequate communications—are just as applicable to any engineering activity as they cover *management* methods. What will be different will be the operational activities and, in fact, we could define a set of operational activities for each of the three major areas we have defined. In doing so we are bound to include some of the operational activities we defined in chapter 2 (eg inspection and test, corrective action and non-conformance control) but others will be rather different. Nevertheless, it does seem, even at first sight, that the concepts of quality assurance enshrined in the existing quality system standards could be applied to these post-installation phases.

Inevitably, though, the detailed interpretation of the standards may wel be different.

We should now look at the "standard" quality system criteria and see which of them are of most relevance or—perhaps more correctly—if any of them are irrelevant to the quality assurance of maintenance and operations. Since BS 5882 (the "nuclear" quality standard) does refer specifically to operational activities it is sensible to consider this standard as well as one of the general standards (BS 5750, NS 5801, ISO 9001 for example). Table 6.2 lists, for convenience, the clauses in BS 5882 and ISO 9001.

Some of the clauses—or at least the topics covered in some of the clauses—are relevant to any activity whereas others are more specifically related to design, manufacturing or inspection. Topics which are *generally* relevant are as follows:

QA programme or system

Organisation or responsibilities

Planning

TABLE 6.2 Clauses of Two Major Quality System Standards

BS 5882	*ISO 9001*
1. QA Programme	4.1 Management Responsibility
2. Organisation	4.2 Quality System
3. Design Control	4.3 Contract Review
4. Procurement Document Control	4.4 Design Control
5. Instructions, Procedures and Drawings	4.5 Document Control
6. Document Control	4.6 Purchasing
7. Control of purchased items and services	4.7 Purchaser supplied Product
8. Identification and control of items	4.8 Product identification and traceability
9. Process Control	4.9 Process Control
10. Inspection and surveillance	4.10 Inspection and Testing
11. Test Control	4.11 Inspection, Measuring and Test equipment
12. Control of Measuring and Test Equipment	4.12 Inspection and Test Status
13. Handling, storage and shipping	4.13 Control of Nonconforming Productivity
14. Inspection, test and operating status	4.14 Corrective Action
15. Non-conforming items	4.15 Handling, Storage, Packaging and Delivery
16. Corrective action	4.16 Quality Records
17. QA records	4.17 Internal Quality Audits
18. Audits	4.18 Training
	4.19 Servicing
	4.20 Statistical Techniques

Review of the quality system, audits

Work instructions or Process Control

Records

Documentation control

Corrective action

Training

If we study the detailed clauses which cover these topics of either of the two standards mentioned (or indeed any other general quality system standard) then we can see that they are relevant to, and can be interpreted in the context of, any maintenance or operational activity. These aforementioned topics set out criteria for organisational control, responsibilities, work instructions and their control, records and their control and training. These, then, are the fundamental quality system criteria.

The other criteria of the quality system standards refer to more specific engineering or manufacturing matters. These will be relevant in some areas.

If we consider maintenance operations (including plant modifications and repairs) then these are engineering tasks which require some inspection and, quite often, purchase of materials and parts, manufacture and perhaps some design work. *All* of the remaining clauses are likely to be relevant.

If we consider "operations" then some of the clauses (or topics) are not relevant. However, topics such as:

Identification of inspection status

Test control

Control of non-conforming items

will be relevant.

It is apparent, therefore, (and this should come as no real surprise) that the quality system approach used in design and manufacture *does* provide an acceptable framework for the quality assurance of operations and maintenance. The interpretation of the criteria may be different and in some cases particular criteria may not be relevant but, aside from that, it is quite appropriate to assess the quality assurance of maintenance and operational activities against the same general quality system standards as used in design and manufacture. That is not to say that specific criteria could not be developed for operations and maintenance in the same way that specific criteria have been developed for the design and manufacture of particular items (eg API Q1 for wellhead equipment). Probably, in the fullness of time, such specific criteria will be generated but that time has not yet come.

MAINTENANCE OF CERTIFICATION

This follows on very naturally from the granting of certification and much of what was discussed in chapter 5 is relevant. After all, under whatever certification regime, the granting of a certificate of fitness or a consent to operate implies that the operator maintains his installation to a sufficient degree that the calculations, reviews, tests and inspections which were carried out to grant the certificate remain valid. In other words, the operator must regularly demonstrate (or have the information to demonstrate when required) that the installation is still as safe as when first approved.

When considering the operator's task in maintaining certification we can think of his actions being of two sorts:

1. The technical actions (structural surveys, testing of safety systems for example).

2. The quality system actions (obtaining assurance that the technical actions have been carried out correctly).

These two sorts of action are analagous to the two sorts of activity in any engineering task (see, for example the discussion in chapter 1 and Figure 1.1).

When constructing a quality assurance system for the maintenance of certification the first step is to define and list the technical actions and their timescale. These will normally be agreed with the certifying authority or, in a self-certification regime, defined to meet the perceived requirements of maintenance of certification. This task is, of course, the prime responsibility of certification engineers who will be aware of the regulatory requirements in all their detail (see chapter 5). This "first step" is some task when we think of the number of certifiable components or systems on a typical platform. In effect, a project plan (for want of a better name) must be worked to if maintenance of certification is to be a smoothly run operation.

Having defined what, in a technical sense, needs to be done then the question of how to obtain assurance of quality can then be addressed. The second step, therefore, should be to construct a quality plan—completely analogous to the quality plans used in design and manufacture (see chapters 2 and 4). This quality plan should be on a number of levels giving different degrees of detail (see Fig. 6.1). Extra complicating factors will be the interfaces between different disciplines and between the operator and specialist subcontractors (eg diving contractors or inspection companies). Ideally, all subcontractors and all specialist groups should operate their own self-contained quality systems so that they can be readily combined into the overall plan—ie the interfaces can be established.

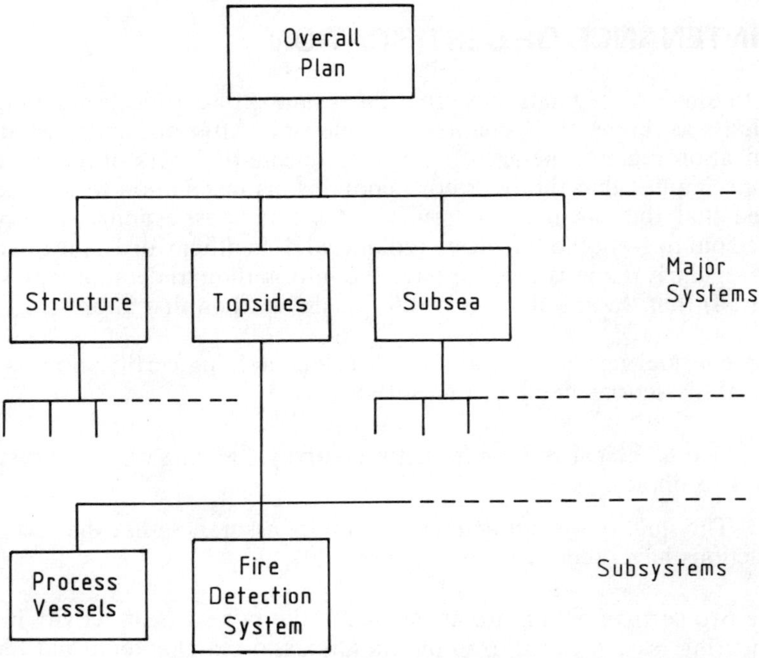

Fig. 6.1 Quality plan for maintenance of certification

The construction of quality plans has been dealt with in earlier chapters (2 and 4) but, briefly, a quality plan must list the operational activities, the controlling procedure or standard, the documentation generated (eg a test record or a certificate of conformance), who is responsible for the action and what review or control of the activity or documentation is required. A quality plan defines the quality activities in a manner which permits them to be "checked off" against the quality requirements.

In the maintenance of certification the emphasis at present is on the generation of authorised documentation or the witnessing or reviewing of activities to demonstrate that the installation is still in a safe condition.

The application of the quality system approach should alter the approach. *If* an activity is controlled by a properly devised quality assurance system, as demonstrated by a sufficiently detailed set of quality plans, then the approach should be to monitor the operation of the system rather than to witness specific operations. This newer approach is logical if we are to apply formal quality management systems to the maintenance of certification. The point about a formal quality system is that it should control an operation and provide evidence that it has been controlled. If effort is to be expended in controlling an operation in this way then the a *logical* way of checking the effectiveness of the control is to audit the system. In other words, as the emphasis on manufacturing control has shifted from final inspection of the product to "inspection" of the quality system the emphasis on a maintenance of certification should shift from documentation review and witnessing of inspections and tests to "inspection" of the quality system.

Certification bodies should therefore primarily be auditing quality systems and agreeing quality plans. This requires quite a different approach than has been common hitherto and therefore has a significant training implication.

This change in attitude will not come overnight but it will surely have to develop the more that operators themselves have in place formal quality systems and do not see the need to duplicate activities.

MAINTENANCE OF THE FACILITIES

Maintenance of the facilities requires, in a way, the same type of actions needed for the maintenance of certification but with a slightly different aim—as was shown earlier in table 6.1. In fact it is sensible to combine the two types of maintenance activity in a single programme which maintains the facilities in an efficient manner and, at the same time, provides assurance to regulatory bodies of the safety of the installation.

Planned maintenance operations are, in reality, the same for an offshore process plant installation as they are for an onshore process plant installation and similar systems of control are valid. The major differences are the necessary use of specialist subcontractors (particularly for subsea work), the cost of subsea work and the management interface between the offshore location and the onshore base. These differences can lead to management and quality problems, the basic communication problems as discussed in chapter 2 and the high cost of lack of control of subsea activities.

The importance of planning should be quite apparent and the meshing together of the complementary quality systems and quality plans of the various participating subcontractors (as in Figure 6.1) is vital. As was shown in other chapters, time and effort spent in quality planning are well spent. This is particularly the case when we consider the very expensive operations (the subsea ones).

Although the aim of maintenance is, as we have stated, slightly different from the aim of maintaining certification it would still be appropriate to use the same quality systems approach. The role of a quality manager in this instance (whatever his exact title may be) should be to monitor the system. This entails constructing the quality plan, ensuring that sub-contractors' quality activities are sufficiently well defined and auditing the outcomes. Again, this requires a significant change in attitude but the important thing to remember is that the approach to be used is really the same as that used in design and manufacture. Why should it be different? After all many of the engineering activities which have to be managed and controlled are the same.

OPERATIONS

Applying a formal quality system approach to the operational activities (ie the production or drilling phases) *is* somewhat different from applying a formal quality system to design and manufacture.

There are two main reasons for this. One is that the *type* of operation is quite different and the demonstration of the assurance of quality must be handled differently. The second is that the timescale of events for control and corrective action is much shorter—production is a continuous operation. As has been very succinctly stated (ref 6.1).

> "The nature of manufacturing activities is such that, in general, time is usually available to allow some form of verification to be carried out without necessarily affecting the quality of the end product, in contrast with plant operational activities, where any deficiency can have a much more immediate effect, particularly with regard to the safety of the plant."

This statement was made with respect to the operation of nuclear fuel plant but is equally applicable to an offshore operation. It is perhaps instructive to note also that only very recently have serious attempts been made to establish *formal* quality systems for the operational activities in industries such as power generation and nuclear fuel re-processing—ie industries where formal quality systems have been applied to the design and manufacturing activities for a very long time. It is self-evident, surely, that such an approach is appropriate to "operations" but it should be recognised that the implementation of this approach to the offshore oil and gas industry may therefore also take quite some time and require a considerable educational activity on the part of operators.

The well documented description (ref 6.2) of the development and implementation of formal quality assurance in the operation of a nuclear power station bears this out.

What we can do at his point is to set out the requirements for a formal quality system and what steps need to be taken to implement one for operations. The requirements and the steps to be taken in fact parallel very closely the situation in other types of activity but as we have stressed before, there are some differences.

Earlier in this chapter we discussed the relevance of the criteria of the quality system standards and established the importance of the following criteria in respect of "operations":

QA programme (or system)
Organisation (or responsibilities)
Planning
Review of the system, audits
Work instructions (or process control)
Records
Documentation and change control
Identification of inspection status
Inspection and test control
Control of non-conforming items
Corrective action

The aim (see table 6.1) is "to provide and implement guidelines, procedures and standards to control the safe operation ..." so it is necessary to look at operational activities in two ways.

Firstly, what *are* the activities and how shall they be carried out and managed in a formal way? Secondly, how are the activities monitored to demonstrate the required assurance of quality?

The core of any quality system for operations must be the operational procedures in all their detail and with all their ramifications relating to fault and emergency situations. These will include, as well as the "conventional" plant operating instructions and procedures, the procedures for coping with emergencies (release of inflammable fluids, fire, explosion, collisions), emergency drills, routine testing of safety systems. Such procedures must include the verification or checking regimes which ensure that operations have been correctly carried out and appropriately authorised. This verification, together with the fault and emergency procedures should be adequate to cope with the time constraint aspect of quality verification.

The bulk of these procedures and instruction will be contained in the various operating manuals relating to different items of plant or different types of operation and, of course, a number of them will need to be reviewed and approved in the context of certification.

Having said this, the problem, as in all occasions when a formal quality system is being developed, is—how far do you go? It is quite easy to take a bureacratic line and prepare "procedures" (all of which need to be controlled documents of course) for every last little step. Clearly this is wrong so, by analogy with other activities, a criticality analysis is required.

HAZOP studies and other reliability assessments (many of which may well be needed for certification purposes) should be the technical means by which operators can judge the significance of particular operations. Such analyses will be the means of making decisions such as:

1. Which operations need documented procedures?
2. How much detail needs to be documented?
3. What are the safe operating parameters of an operation?

3 will lead to considerations of safety and emergency situations and the need for (and detail of) procedures for such situations. This stage (criticality analysis leading to definition of procedures) is analagous to defining the "technical level of quality" (see chapter 1) and it is a matter for the operations staff, taking due note of certification requirements. The aim should be to identify the critical processes and control and document them. As has been said earlier, attempting to document a procedure is a very good discipline in itself as it focuses the thoughts on what has to be done, who has to do it and to what degree of precision it needs to be done. Quite often these matters have not been thought of logically in devising operating policies.

The devising and documenting of the operational procedures will, therefore, meet the following quality system criteria:

Work instructions (obviously)	
Control of non-conforming items Corrective action	Emergency and fault procedures
Test control Identification of inspection status	Should be inherent in the operating procedures

The rest of the task, then, is the monitoring of the management of the defined operational activities to demonstrate the assurance of quality. This means satisfying the following quality system criteria:

QA programme (or system)

Organisation

Planning

Review of the system, audits

Records

Documentation and change control

Training

These criteria are the "true" quality management criteria and they relate to the "people" aspects, the interfaces, the information and reporting methods and, not least, the training in the concept of a quality system approach. The problems in these aspects relate to the time constraints imposed by the fact that operations are continuous and there is no time (or not enough time) to stand back and carry out quality verification in a relatively relaxed way. Operations staff are obviously well aware of these time constraints and will not willingly participate in extra form filling merely to satisy the "quality" people. Hence the extreme importance of training to make people aware of the importance of quality assurance and the need, also, to approach quality assurance in operations in a step-by-step way. In other words, apply the principles gradually starting with a few key operations and then, as people see their worth, apply them more widely. One clear requirement of a formal quality system is more clearly defined accountability for actions. Operating staff will need to carry out more formal checks and "sign off" activities more frequently. This is not a bad thing at all, provided it is done sensibly, as signing for something generally makes a person take great care that they have done it properly. This "signing off" should be the main and most obvious change for the majority of operating staff—it is the management who will principally be affected by changes. It is the management who, after all, will control the documentation, carry out the system reviews, and be responsible for keeping correct records and acting on any corrective action needed.

Obviously, someone will have to have the responsibility for monitoring the quality system (ie the "management representative"), there will need to be a defined organisation with "quality" responsibilies defined and, of course, a quality manual to define and document the overall policies. All of these should be directly analogous to the situation in design and manufacturing—as should the documentation control and internal review (and audit). Again, all the benefits (and problems) which arise in applying formal quality systems in other areas will also apply here.

CONCLUDING REMARKS

The very limited application, to date, of formal quality systems to maintenance and operation makes it difficult to give precise guidance as there is little practical experience of the problems which will arise.

What we can say is that the philosophy is just as applicable to these areas although there will be differences in emphasis. The important things to remember are the five major features of a quality system and the necessity of defining as precisely as practicable the technical or engineering requirements. Otherwise the approach to applying a formal system (and even the available quality system standards) is just the same as in the more traditional areas.

The main changes will probably need to be in attitudes (the attitudes of certification engineers and operating staff for example) which implies a training and familiarisation programme and also implies that formal systems will take some while to become commonplace in the industry.

REFERENCES

6.1 Derbyshire, W. Applying QA to Nuclear Plant Operations. *Quality Assurance in Plant Operations.* The Institute of Quality Assurance. London 1986.

6.2 Lake, M. W., Steele, K. and Jenkins, G. Recent Developments in Operational QA within the CEEB. *Quality Assurance in Plant Operation.* The Institute of Quality Assurance. London 1986.

CHAPTER 7

THE COST OF QUALITY ASSURANCE ACTIVITIES

E. M. Rooney

INTRODUCTION

In the preceding chapters the various aspects of quality assurance, as applied to the offshore oil and gas industry, have been discussed. In all cases, the point has been made (or at least implied) that measures to demonstrate the assurance of quality are labour intensive; ie they need people to define and document procedures, to monitor performance, to review records, to analyse deficiencies and to carry out corrective actions and so on. In turn, the quality control activities, which are always necessary to some degree to measure and evaluate quality levels, are themselves labour intensive.

The nature of all this activity necessarily, therefore, implies a cost. We take it as read that the extra cost of implementing a formal quality assurance programme must be justified by the benefits in improved quality leading to better project completion times, more efficiently run projects and greater reliability of the equipment and facilities once in service. This, after all, is the basic justification for formal quality systems, ie there must be a cost/benefit trade off.

In the earlier chapters we have concentrated, primarily, on the benefits of quality assurance programmes without considering in any detailed way the costs incurred. It is, of course, obvious that the more detailed and rigorous the quality assurance programme, the more costly it will be to implement and a point will always arise when the cost of an "extra degree" of assurance is not justifiable in terms of a reduction in probability or consequence of failure. This is the reason, of course, for defining different levels of quality system in the recognised quality system standards (see chapter 2) and for the use of criticality ratings (particularly at the design and specification stage—see chapter 3). However, these are relatively crude measures for defining the required level of quality programme and it would be helpful if the cost of implementing formal quality programmes was known with more precision.

Many analyses are made of the benefits of quality assurance (including the various systematic risk and hazard analysis techniques and reliability assessments) but similarly rigorous analyses are rarely made of the costs.

One reason for this lack of quality programme cost analysis is often given as the difficulty of defining what is a true "quality" cost and the fact that different organisations (and projects) operate different internal costing systems. What this chapter does is analyse the "quality" costs in the offshore industry primarily by illustrating how quality related costs *can* be effectively measured in two quite different types of offshore project and, hence, giving guidance in a more general way on how to identify, measure and, hence, *control* the quality costs in any project.

The requirement for formal quality assurance schemes and activities by clients and contractors in the offshore industry is now commonplace. Such schemes are recognised as necessary to reduce the probability of expensive failures of high cost, high integrity plant to a negligible level. Coupled with this is the requirement, in some circumstances, for certification (as discussed in chapter 5). Operators then need to demonstrate that items for offshore installation and use meet certain statutory requirements and a quality assurance approach can help demonstrate that requirements of safety and fitness for purpose have been achieved. Historically, there has been a pattern in the offshore industry (and process industries generally) of quality assurance and quality control activities being closely monitored on individual projects by auditors, third party inspection agencies and certifying authority inspectors. This was highlighted by Plunkett (ref 7.1), in a survey of process plant fabricators. His work also highlighted the fact that although, in many instances, quality assurance and quality control activities were recognised as significant factors in the design and production process, few companies measured or had any idea of the cost of their quality assurance operation.

In the offshore industry, contracts are of a very "project specific" nature. Every contract has its own particular requirements and, therefore, its own particular QA requirements. These are ultimately determined by the client operators who, of course, are ultimately responsible for the reliability and efficiency of equipment installed offshore and for meeting certification requirements where applicable. The costs of quality activities are, therefore, likely to vary from project to project as they are inevitably dependent on the particular QA requirements.

WHY MEASURE THE COST OF QUALITY ACTIVITIES?

The reasons for measuring "quality" costs derive from the fact that quality assurance is a management activity and quality costs are a useful tool for decision making. This was shown by Feigenbaum (ref 7.2). Quality costs are the "tools" whereby decisions can be made that affect costs in various areas in order to obtain minimum quality cost at the desired level of quality and of assurance of quality. For example, various applications of quality cost measurement were considered by Feigenbaum:

Quality costs as a measurement tool
Measurement can help justify investment—eg to reduce failure costs or reduce inspection costs.

Quality costs as a process-quality analysis tool
When quality costs are broken down by area of the engineering activity, this helps pinpoint problem areas—eg design or material control.

Quality costs as a programming tool
Quality costs can be a means for identifying actions which provide the highest potential payoff in terms of improving efficiency. Hence these actions can be given priority.

Quality costs as a budgeting tool
Quality costs are a guide to budgeting for the necessary expenditure for accomplishing required QA programmes.

Quality costs as a predictive tool
Cost data provide the controls to evaluate and assure performance in relation to the goals and objectives of the company. Known quality costs are effective in producing valid cost estimates in obtaining new business.

All these management aspects are relevant and important in the offshore industry as will be shown.

An efficient, effective quality assurance system implies efficient, effective management. Efficient management depends on efficient allocation and control of resources. Resources may be measured in various ways, and the two most useful parameters for this labour intensive activity are in terms of man hours or in terms of actual cost. To know and to understand the efficiency of QA operations at any level (client, contractor, manufacturer), the resources devoted to those operations need to be measured and monitored. Remember, the major cost of any "quality" activity is the labour cost—not the cost of the hardware.

HOW QUALITY RELATED COSTS ARE DEFINED

There is a British Standard (ref 7.3) which gives some guidance on quality costs and how to allocate them. Quality costs are defined in this standard (BS 6143) as follows:

Prevention costs: the costs of any action taken to investigate, prevent or reduce defects and failures.

Appraisal costs: the costs of assessing the quality achieved.

Internal failure costs: the costs arising within the manufacturing organisation of failure to achieve quality specified (before the transfer of ownership to the customer).

External failure costs: the costs arising outside the manufacturing organisation of failure to achieve quality specified (after the transfer of ownership to the customer).

This standard suggests a format for cost reporting of quality activities with a view to carrying out trend analyses. In the offshore industry, the contractual nature of the business makes ongoing quality cost analysis in this way virtually impossible in most cases, so a different approach is required for quality cost analysis.

This is not to devalue BS 6143 in any way. It is an excellent standard for application to batch or mass production type industries where identical (or almost identical) product is made on a continuous basis to effectively the same standards of quality and assurance of quality. This is very rarely the situation in the offshore industry (even in the supply of proprietary items like valves).

An alternative approach to the categorisation of quality related costs was devised by Moyes and Rogerson (ref 7.4) specifically for the process plant industry. As this is of the same nature as the offshore industry (in terms of the contractual nature of the projects) this categorisation is just as applicable to the offshore industry.

Briefly, the approach is to concentrate on the first three categories of cost defined in BS6143 and ignore "external failure costs" as they are so large (and variable) that no real value is obtained from measuring them (at least in the context of a management tool for optimising the efficiency of a quality programme). The very size of external failure costs is after all the *reason* for formal quality programmes in this industry!

Next, quality activities are divided into "system costs" (the cost of the basic, minimum quality system appropriate to the industry) and "project specific" costs (the quality costs which are *directly* attributable to a particular project). It is these latter costs which are the variable ones and which most need measurement and control.

The major headings under this latter category are:

1. Auditing by clients and/or contractors
2. Auditing of suppliers and subcontractors
3. Inspection and control of bought-in items
4. Production and approval of special procedures (particularly welding and NDT)

5. Special operator approvals (particularly welding and NDT)
6. Quality engineering
7. Validation inspection
8. Repair and rectification
9. "Waiting time" and other quality related delays

These have been found useful in allocating quality costs in particular areas of design and manufacturer activity. These categories can also readily be used for cost measurement in offshore projects and, later on in this chapter, two different types of offshore project are analysed in this way and the applicability of these cost categories is demonstrated.

HOW QUALITY RELATED COSTS ARE MEASURED

In spite of the importance of the subject and the general belief (particularly among manufacturers and fabricators) that costs associated with quality assurance programmes are higher than they should be, there are very little reliable, quantitative data which are relevant.

Hicks (ref 7.6) examined the cost effectiveness of QA programmes for the steel fabrication industry, particularly offshore oil and gas. He touched on the dangers of overspend on QA programmes, especially with regard to a client's natural wish to monitor a fabricator's QA activities. Clients should not duplicate the QA function—an inability to monitor is analagous to an inability to delegate. Hicks succinctly expresses the problems of multiple inspections and imposed QA requirements and this point is considered in more detail in one of the examples studied here.

Quality costs are not hard to measure. Even in situations where QA is not formally costed, the costs can be arrived at with relative ease. Company financial records tend to be adequate for costing QA activities, and man hours, where logged or documented in some way, are very useful. For specific one-off projects some estimation of man hour input may be necessary but this will be shown to be possible with sufficient accuracy to give a reasonably successful result.

Cox (ref 7.3) pointed out that the problem of measuring and controlling the cost of quality is not unlike any other management and management accountant's problem. To analyse quality costs, quality activities need to be separated out from other activities and costed separately, primarily in terms of their man hour requirement at a realistic man hour rate. Hence, quality costing is a two part exercise and should be a joint effort by the quality assurance manager and the management accountant.

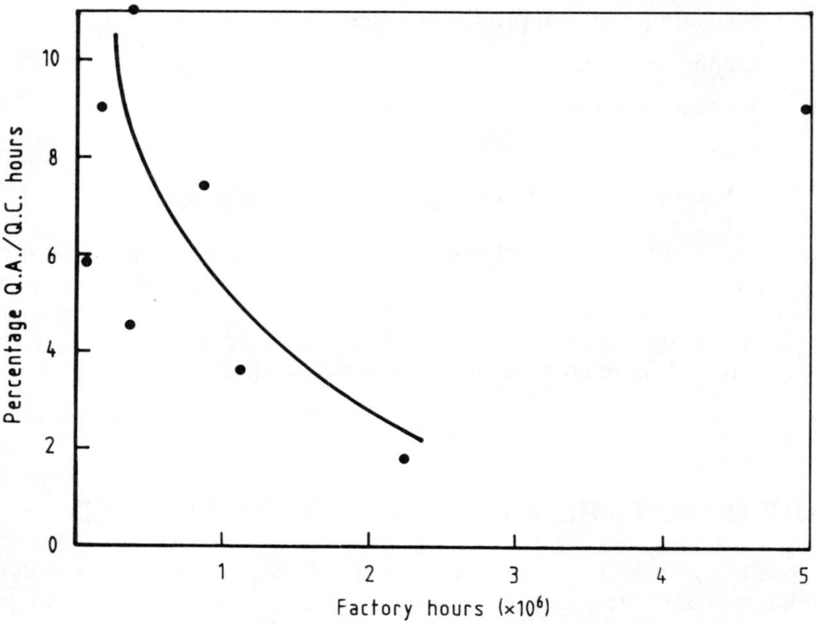

Fig. 7.1 Quality system costs in terms of the proportion of total factory man hours devoted to QA and QC activities for eight UK process plant manufacturers

Much has been written on the subject of quality assurance costs but there has been a noticeable lack of data in this important area of management responsibility. Quality costs of themselves are not a useful tool unless measured in some index. As mentioned earlier, "system costs" are one suggested measure of quality costs. Figure 7.1 shows the annual quality system costs of eight UK process plant manufacturers. Note that these costs have been expressed in terms of the proportion of annual QA and QC man hours in relation to the factory's overall annual man hours.

In the process plant industry, Moyes and Rogerson (ref 7.4) found contract selling price, or contract selling price less subcontracted costs, a useful index and Figure 7.2 indicates, from this reference (7.4), quality related costs of some process plant manufacturers expressed in this way.

Two examples will now be examined. In neither case were quality related costs initially formally recorded or documented but, in each case, as we will demonstrate, it was possible to measure them without too much difficulty by scrutiny of appropriate engineering and financial records.

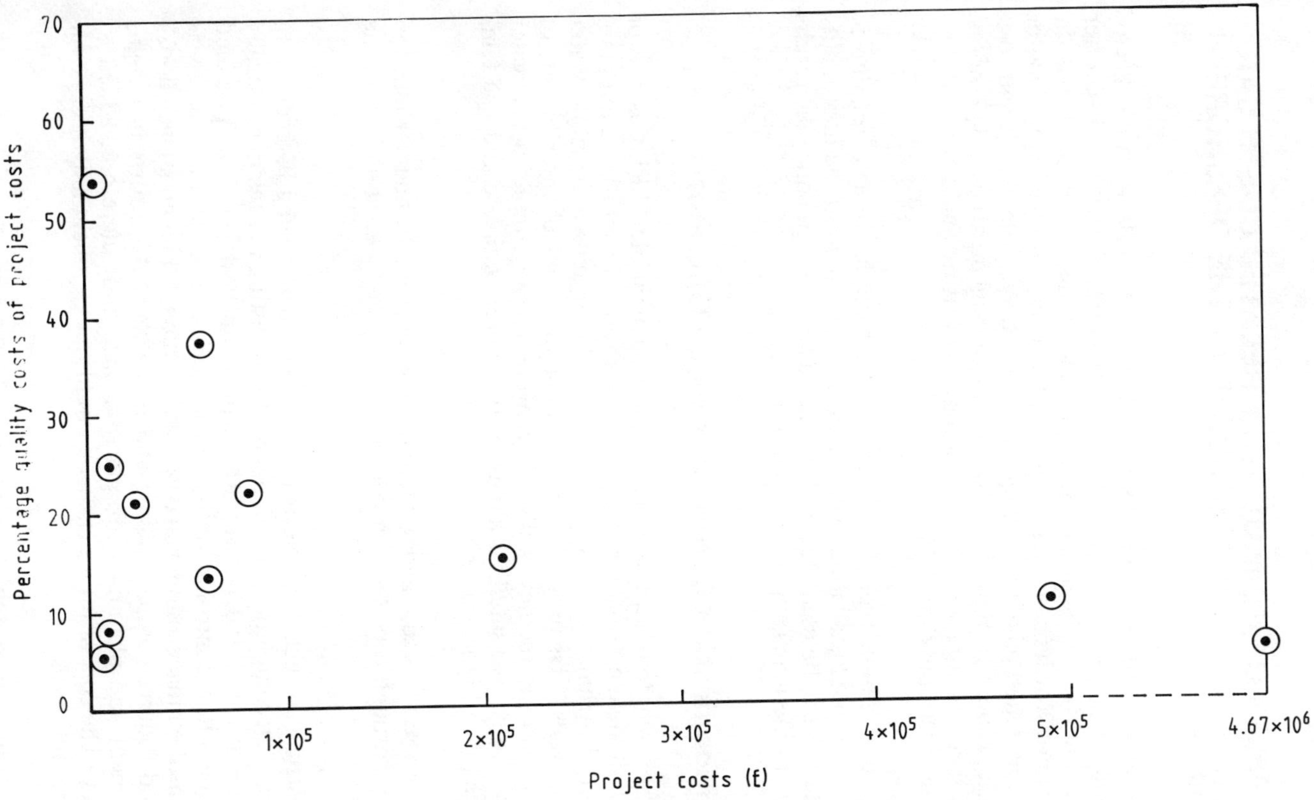

Fig. 7.2 Project quality related costs in terms of QA and QC costs in relation to project costs for 11 process plant items. (Source: ref. 7.4)

EXAMPLES OF HOW QUALITY RELATED COSTS CAN BE IDENTIFIED AND MEASURED IN THE OFFSHORE INDUSTRY

The evidence is that, in most cases, quality related costs are not collected and analysed in any formal way in the offshore industry and it is often thought that the problems of measuring such costs are so great that the effort is not worthwhile. This is not so, however, *provided* you view the exercise as a method of identifying the relative magnitude of different costs and their proportion of total project costs for management control purposes, rather than as a pure accountancy exercise to measure quality related costs to the nearest pound or dollar.

Two quite different types of offshore project (both, though, relating to manufacture and installation) will now be described to illustrate how quality related costs can be identified and measured and what guidance the results can give to the management of such projects.

Field Modification Carried out by an Operator

This first example relates to a subsea injection manifold which was part of an offshore field extension project. The extension of water injection pipework and instrumentation was for a further four subsea template water injection wells. This provided a means of supply, control and metering of injection water to the new wells. There were, already, three subsea water injection wells, two satellites adjacent to a subsea template and one template well (see Figure 7.3).

Injection water was supplied to these three wells from the main production platform through pipework, a composite riser and two in-field pipeline bundles.

The design was such that the whole system did not need a recertifying hydrotest. On completion of well tie-ins, hydrotesting was restricted locally to the tie-in spool and not the whole injection flow leg. This provided independent flow control at all the new locations. A manifold concept was chosen as the most economic and operable option as individual well hook-up would require complex piping and be costlier. All efforts in design, procurement and planning took place at once and, therefore, added to efficiency. The cost of future changes would rise exponentiallly if done on a well-by-well basis.

All quality assurance activities were carried out by the client company, in-house. Specifications, inspection and certification were performed by the appropriate departments. Physical testing, such as hydrotesting, was carried out by subcontractors.

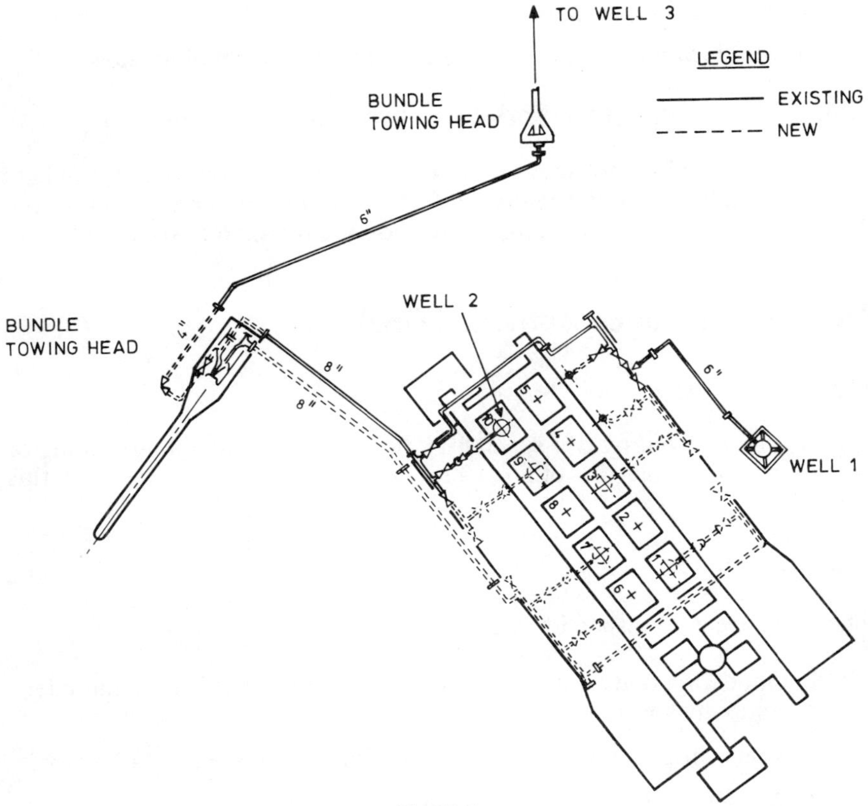

Fig. 7.3

This is a useful example to study since there is a feeling on the part of manufacturers that clients have a tendency to impose excessive or unnecessary quality requirements of which fabricators have to carry the costs. The client company studied, a North Sea operator, had an "inspection oriented" approach to quality assurance. His policy was that since the operator is responsible for the efficiency, quality and reliability of his equipment, the operator should take full responsibility for all aspects of quality assurance connected with the project. This is, perhaps, a slightly unusual situation from a fabricator's point of view (and certainly very different from the other example described).

The costs outlined cover design, procurement, manufacture, fabrication, installation and commissioning of the subsea manifold and the related topside control and instrumentation.

Three fabricators were appointed to carry out fabrication onshore of:

1. Spool protectors and pipe support

2. Pipespools and spool transport frames
3. Modification of the handling frame and new handling aids

This onshore phase of the work took approximately four months in total.

Offshore, one fabricator was appointed to do pipespool and structural field fit and completion, and one diving contractor was appointed to carry out the subsea work. This offshore phase took approximately six weeks.

Quality Assurance Activities Onshore

Hydrotesting

This was carried out by two specialists, one of whom carried out testing of the pipespools while the other acted as an independent witness of this procedure.

Inspection

The main areas of inspection were:

1. Goods in/goods out inspection of free issue material at the client company's warehouse.
2. Visual inspection by an appointed third party at each fabricator.
3. Welder approval mechanical tests.

Non-destructive Testing

NDT carried out at fabricators was:

1. X-ray inspection of welding.
2. Ultrasonic and magnetic particle inspection.

Each of these activities was carried out by specialist NDT contractors appointed by the client company.

Quality Assurance Activities Offshore

Hydrotesting

Again, this was carried out by two specialist contractors, one of whom carried out on deck and subsea hydrostatic testing while the other was an independent witness to this.

Inspection

Leak detection of flexibles and flushing of umbilicals was carried out by a third party inspection agency.

Non-destructive Testing

Some X-ray inspection of welds and some ultrasonic inspection was carried out offshore by NDT contractors.

In addition to these onshore and offshore activities, there was QA input from the Quality Assurance Manager, who selected the quality assurance subcontractors and scheduled and monitored their work. The project was also subject to certifying authority approval, the costs of which were charged directly to the client.

The only QA activity carried out by the operator in-house was the quality engineering. This was the responsibility of the quality assurance manager. A large part of his work involved selecting, approving and monitoring the work of subcontracted QA/QC specialists.

Where timesheets and invoiced costs were readily available, man hours and overall costs could be readily obtained. In some instances, not all the subcontracted QA activities were in fact charged directly to the project, and these had to be extracted from timesheets, monthly payments, offshore allowances etc.

All quality cost data were documented in some form and could be retrieved. There were very few instances of missing data, and where this did occur, it was possible to make an estimate, because the cost of all items and services provided for this project were well documented and it was possible to highlight the QA related activities and note their costs. It is a feature of most offshore related work that very detailed quality records are available (if only to meet the needs of certification) so better estimates of quality costs can be made than for many onshore projects. The costs of all subcontracted work charged *directly* to this project were well recorded and documented.

The findings are summarised in table 7.1. It can be seen immediately that the *documented* quality related costs were in fact inspection, test and certifying costs.

The overall quality costs as a proportion of *project* cost were:

NDT and inspection	12.74%
Mechanical testing	0.20%
Certification	0.21%

TABLE 7.1. Field Modification Project: Quality Related Activity Costs Charged *Directly* to the Project

DESCRIPTION	*QA ACTIVITY*	*QA ACTIVITY COSTS AS % OF TOTAL PROJECT COSTS*	*QA ACTIVITY COSTS AS % QA ACTIVITY*
Onshore Services	NDT, Ultrasonics	1.25	9.53
	X-ray inspection	0.90	6.83
	Mechanical testing	0.20	1.55
	Hydrotest services	2.35	17.88
	Hydrotest witnessing	3.22	24.48
Offshore Construction Topside	NDT, Ultrasonics	0.90	6.84
	X-ray inspection	0.62	4.70
	Hydrotest services	1.79	13.63
	Hydrotest witnessing	1.71	12.99
Certification		0.21	1.57
GRAND TOTAL		13.15	100.00

and

individual quality item costs as a proportion of *total quality* costs were:

NDT and inspection	96.88%
Mechanical testing	1.55%
Certification	1.57%

Indirect Costs

These figures above were not, however, the total picture in quality assurance terms. There were other quality activities for the project which were not invoiced or directly charged in any way, but are nevertheless part of the total quality effort.

Quality engineering, which amounts to the input of the QA manager, was not included in the direct costing.

Goods received inspection of free issue material at the client company's warehouse was not costed directly to the project.

Some *visual inspection* carried out on and offshore was also not charged directly to the project.

The inspection costs were available in the form of timesheets, which were examined. The time devoted to quality engineering was estimated by the quality assurance manager on a realistic basis.

As the monthly costs/charges of the personnel concerned were known from company accounting records, a "true" hourly rate could be estimated.

These man hour rates were then applied to the hours devoted to the activities of quality engineering, goods received inspection and visual inspection.

A summary of the total quality related costs of the project, which includes these indirect costs and details of man hour input for each QA activity, could then be constructed and is shown in table 7.2.

Examination of Quality Costs

It can be seen, immediately, that the costliest item is hydrostatic testing and that the cost of this is more than doubled by having this witnessed (29.73% of quality costs devoted to the hydrotest; 65.11% devoted to hydrotest plus witnessing). Next, in order of magnitude, are ultrasonic inspection and radiography (15.46% and 10.88% respectively). The costs of visual inspection and quality engineering are very low (2.88% and 2.20% respectively). Special approvals, certification and goods-in inspection rank so low, as to be almost negligible (1.46%, 1.48% and 0.53% respectively).

The costs of *indirect* activities in relation to total quality costs are as follows:

Quality engineering	2.20%
Goods received inspection	0.53%
Visual inspection	2.88%
TOTAL	5.61%

The fact that these costs are *not* charged directly to the contract is not important, since their absence does not grossly distort the figures (see below). The substantial costs such as hydrotest and NDT are well recorded.

Examination of Quality Costs on a Man Hour Basis

If the above quality related costs are looked at purely from man hour effort, it is immediately obvious, and not surprising, that hydrotesting is still the biggest quality related input. Of all the *time* devoted to the quality related activities, over 70% is accounted for by hydrotesting. NDT accounts for just over 16% of the effort. Of the remaining tasks, quality engineering and visual inspection, *together* account for just over 8% of quality hours.

Obviously, examination of table 7.2 shows that where differences occur in quality costs as measured in terms of man hour input or in terms of total financial costs, this has arisen because of the additional costs such as equipment hire, mileage and offshore allowances. Examination of table 7.3, based on the Moyes and Rogerson activity headings, indicates that whatever basis is used to measure the quality costs of the project, the findings are similar in that the relative magnitudes of the costs with respect of each other do not vary very much.Looking at the total quality related

TABLE 7.2. Field Modification Project: Quality Related Activity Tasks and Costs, Both Direct and Indirect Costs.

Note: This table includes the direct costs from Table 7.1

QA ACTIVITY	*TASK*	*MAN HOURS DEVOTED TO THIS ACTIVITY*	**ADDITIONAL COSTS (£)*	*% MAN HOURS OF TOTAL QA MAN HOURS*	*%COSTS † OF TOTAL QA COSTS*	*COMMENTS*
Quality Engineering	Input by quality manager: meetings, negotiations, reports, appointing and monitoring subcontractors.	160	—	4.81	2.20	Estimated from discussions with the quality manager.
Visual Inspection	Inspection at two fabricators and leak detection carried out offshore.	119	119.97 (offshore allowance)	3.58	2.88	These figures were arrived at from examination of time sheets, known monthly rates and known offshore allowances.
Goods in/ goods out	Inspection carried out at warehouse of free issue material.	25	33.00 (mileage)	0.75	0.53	These figures were acquired from timesheets and known monthly rates.
Hydrostatic pressure testing	Hydrotesting of spools on and offshore.	1358	6724.37 (equipment hire)	40.88	29.73	These figures were obtained from time sheets and invoices.

Witnessing of pressure tests	Witness of above.	1017		30.61	35.38	These figures were obtained from time sheets and company's project figures.
Radiography	Inspection carried out at two fabricators and offshore.	316	89.75 (mileage)	9.51	10.88	Figures obtained from time sheets, invoices and project accounts.
Ultrasonic and MPI inspection	As above.	234	9234.37 (mileage plus equipment hire)	7.04	15.46	Figures arrived at from time sheets, known monthly rates and project accounts.
Special Approvals	Mechanical and non-destructive testing of welder approval test pieces.	44	—	1.32	1.46	Figures arrived at from project accounts and invoices.
Certification	Input of certifying authority's representative. This is a statutory requirement.	50	—	1.50	1.48	The 50 man hours are an estimate made by the author, based on project accounts figures.
GRAND TOTALS		3323		100.00	100.00	

*Additional are those costs in addition to man hour costs, e.g. travel expenses, equipment hire, offshore allowances.
†These costs include the cost of the man hour inputs plus any additional costs that were incurred.

TABLE 7.3. Field Modification Project: Summary of Quality Related Costs and the Comparison between Inclusion and Exclusion of *Indirect* Costs.

QA ACTIVITY	*COSTS AS % OF DIRECT QA COSTS RECORDED IN PROJECT ACCOUNTS*	*COSTS AS % OF TOTAL QA COSTS (IE DIRECT AND INDIRECT QA COSTS)*	*HOURS AS % OF PROJECT QA HOURS*	*DIRECT COSTS AS % OF PROJECT COSTS (IE DIRECT QA COSTS RECORDED IN PROJECT ACCOUNTS)*	*DIRECT AND INDIRECT COSTS AS % OF PROJECT COSTS*
Validation Inspection	96.88	94.33	91.62	12.74	13.14
Quality Engineering	—	2.20	4.81	—	0.30
Certification	1.57	1.48	1.50	0.21	0.21
Special Approvals	1.55	1.46	1.32	0.20	0.20
Goods in/ goods out	—	0.53	0.75	—	0.07
TOTALS	100.00	100.00	100.00	13.15	13.93

costs with respect to project costs, there is really little difference if these are analysed from the project accounts or the more comprehensive analysis. The difference is 0.78% and this suggests that as far as monitoring quality costs of a project such as this, the project accounting technique used is adequate.

If a "full" analysis is required, such that the indirect costs quality engineering, visual inspection and goods-in inspection are included, this could be adequately done by logging man hour input throughout the project.

Summary of Quality Cost Findings

The quality cost findings are shown in table 7.3 under the headings outlined earlier. Quality assurance in this specific instance amounts to *validation inspection* (visual, NDT, hydrotest). The operator's overall approach to assuring quality was to take responsibility for inspection. For practical and budgeting purposes, it makes little difference whether the company's project accounts (ie direct charged QA costs only) are used to assess these inspection costs or the more comprehensive approach (taking account of indirect costs also). The difference is 2.55% of total quality costs *or* 0.4% of project costs.

The remaining QA activities amount to less than 6% of total quality costs or less than 1% of project costs.

It makes no significant difference whether quality costs are measured in terms of finance (column 3 of table 7.3) or of QA hours (column 4 of table 7.3). Activities aside from validation inspection contribute little to quality costs whether these are presented in terms of cost or man hours.

It can be seen, therefore, that identifying and measuring the quality related costs in a project organised in this way is a relatively straightforward procedure. The amount of time needed to do such an analysis is, of course, dependent on the degree of detail required in the figures. In general, the approach should be to do a "quick" analysis to get order of magnitude figures and then do more detailed analyses of areas which are particularly important (in this case the cost of hydrotesting). Any management action to improve the efficiency of the quality assurance in a project organised in this particular way should be directed at improving (if possible) the efficiency of the hydrotest stage. However, as we shall see later, this method of organising a project implies certain assumptions about the competence and experience of the operator in selecting suppliers and subcontractors.

Manufacture of a Pressure Vessel for Installation on a Producing Platform

This example relates to the "traditional" construction phase of a project where the order for a vessel has been placed with a small fabricator (120 employees) by a major contractor acting on behalf of an offshore operator. The vessel was a conventional carbon–manganese steel vessel, of 14000 kg fabricated as a category 1 vessel to BS 5500 (1982) and the contract price was £30000. Since the vessel was for use on an offshore installation, it was subject to certifying requirements and this took the form of inspection carried out by a certifying authority representative.

The fabricator was under contract to the contractor who had appointed the third party inspection agency. Fabrication of the vessel took 12 weeks. (The cost figures quoted relate to 1982 prices).

Unlike the other example, the fabricator was expected to be responsible for his own quality assurance. This was how the company normally operated. The company was expected to have a quality system to BS 5750 Part 1 for this offshore contract, but in fact the company normally operated to BS 5750 Part 2. The fabricator had no in-house NDT or mechanical testing facilities, so all of this was subcontracted to testing specialists and the fabricator was totally responsible for this and carried all the costs of subcontracted quality assurance. They frequently did work for the fabricator who was confident of their skills and ability.

There were two in-house, shop-floor inspectors who carried out visual checks and monitored NDT activity and records. In addition to this, the contractor appointed a third party inspection agency and their representative paid frequent visits to the fabricator's premises to carry out his own, independent inspection. The cost of this was borne by the contractor. A certifying authority inspector also visited at intervals and usually wished to inspect the same items that the third party inspector had checked. This is a good example of the much complained of multiple inspection situation. In fact, in this instance, multiple inspection was not considered a great problem but some delays, resulting from awaiting inspection clearance, were encountered.

Quality Assurance Activities

In order to win the contract, the fabricator was expected to demonstrate a quality system to BS 5750 Part 1. The company was audited by the client. The audit was, in fact, a brief, half day survey carried out at the fabricator's premises.

Within the fabricator's premises, the quality assurance input came from several people. There was a technical director who spent 20% of his time in a quality assurance manager's role. Documentation was controlled by a technical clerk who monitored all paperwork connected with quality assurance, including subcontractor's invoices and checked all documents and material certificates connected with goods receiving. There were two shopfloor inspectors who carried out all necessary visual inspection and dimensional checking that was required on the vessel. These inspectors accompanied the third party inspector and the certifying authority inspector when they came to examine the vessel, check radiography or examine other NDT reports.

All NDT and mechanical testing (of production weld test plates) was carried out by subcontractors. The fabricators had no in-house test facilities of this nature. The direct costs of NDT to the fabricator were readily available from the appropriate invoices.

Measurement of the Costs Incurred

The quality related costs for this project were borne primarily by the fabricator and can be classified under the headings indicated earlier. As for the previous example, the fabricator did not keep formal records of quality assurance and quality control costs but all the subcontracted NDT and mechanical testing costs were directly available from invoices. The rest of the data was in the form of time input by the fabricator's personnel which could be established without much difficulty.

1. Auditing by Clients

As mentioned before, this amounted to a half day visit to the fabricator by a *client* (offshore operator) representative. The fabricators' main participant here was the technical director.

2. Auditing of Suppliers

No work was done under this heading.

3. Inspection and Control of Bought-in Items

This was done by a technical clerk who checked documents and material certificates of delivered material.

4. Production and Approval of Special Procedures

The special procedure requirement was very high in this project. Welding procedure tests had to be carried out and the test plates tested by subcontractors.

5. Special Operator Approvals

No project specific demands were made in this area.

6. Quality Engineering

Quality engineering was the responsibility of the technical director and for this project it amounted to attendance at a pre-order meeting held at the contractor's premises and various inspection meetings held at the fabricator's premises.

7. Validation Inspection

Visual inspection
This was carried out by the two inspectors on the fabrication shop floor.

Production welding test plates
These were required by the contractor and they were mechanically tested and radiographed by subcontractors.

Non-destructive testing
Radiography, magnetic particle inspection and ultrasonic inspection were all required and all carried out by subcontractors appointed by the fabrication company.

Hydrostatic testing
The hydrostatic pressure test of the completed vessel was carried out by the fabricator.

8. Repair and Rectification

Some cracking was detected in welded joints so it was necessary for the fabricator to grind out, re-weld and re-inspect by radiography.

9. Waiting Time and Other Quality Related Delays

In situations of third party inspection and certifying inspection, there is much complaint about time spent waiting for items to be inspected at inspection "hold points". Some waiting time was experienced by the fabricator in this instance, but it did not prove to be a significant problem.

Examination of Quality Related Costs

The quality costs of the vessel are summarised in table 7.4. The time spent was costed from the average industrial man hour rate.

The work carried out by Moyes and Rogerson (ref 7.4) showed that irrespective of fabricators' size and complexity, true man hours in this industry are surprisingly consistent.

Calculated true man hour rates were confirmed as accurate by cross check-

TABLE 7.4. Pressure Vessel Contract: Quality Related Tasks and Costs

QA ACTIVITY	*TASK*	*MAN HOURS DEVOTED TO THIS ACTIVITY*	*COST OF MAN HOURS AT £6.71* PER HOUR (£)*	*INVOICED AND ADDITIONAL COSTS (£)*	*TOTAL COSTS (£)*
Auditing by the client	Assessment of quality system. Technical director in full attendance. Other personnel as required.	12	80.52	—	80.52
Goods-in inspection	Identification and checking of certificates.	36	241.56	—	241.56
Special procedures	Welding procedure test plate production.	100	671.00		
	Mechanical testing of plate.			850.00	
	X-ray of plate.			8.00	1529.00
Quality engineering	Pre-order meeting and inspection meetings attended by technical director.	20	134.20	50.00 (travel)	184.20
Validation inspection	Visual inspection.	58	389.18		
	Production test plates.	40	268.40		
	Mechanical testing of test plates.			1015.00	
	X-ray of test plates.			11.00	
	MPI testing.			50.00	
	Radiography.			1344.00	
	Ultrasonic testing.			183.00	
	Hydrostatic testing.	48	322.08		3582.66
Repair and rectification	Weld repair.	133	892.43		
	Re-radiography.			106.00	998.43
Waiting time	Waiting for inspection.	120	805.20		805.20
TOTALS		567	3804.57	3617.00	7421.57

*Man hour rate estimated from industry average for 1982.

ing with data from the census of production developed by the Business Statistics Office in the UK (similar data are available in other countries).

Therefore, for the example studied, the average man hour rate relevant to the industry for year of manufacture (1982) was applied to the man hour figures as shown in table 7.4.

From table 7.5 it can be seen that the quality related costs as a proportion of the contract price are 24.74%. In other words, approximately one quarter of the selling price of the vessel was devoted to quality assurance and

TABLE 7.5. Pressure Vessel Contract: Summary of Quality Related Costs

QA ACTIVITY	MAN HOURS DEVOTED TO ACTIVITY	TOTAL COSTS OF ACTIVITY (£)	% MAN HOURS OF TOTAL QA* HOURS	% COSTS OF TOTAL QA COSTS	% COSTS OF PROJECT COSTS
Auditing by client	12	80.52	2.12	1.08	0.27
Goods-in inspection	36	241.56	6.35	3.25	0.81
Special procedures	100	1529.00	17.64	20.61	5.10
Quality engineering	20	184.20	3.53	2.48	0.61
Validation inspection	146	3582.66	25.74	48.28	11.94
Repair and rectification	133	998.43	23.46	13.45	3.33
Waiting time	120	805.20	21.16	10.85	2.68
TOTALS	567	7421.57	100.00	100.00	24.74

*These hours do *not* include the hours of subcontracted personnel.

quality control activities undertaken by the fabricator. This is actually well within the range which has been found for the process industry generally in the study carried out for NEDO (7.4) in which quality related costs were in the range 2.50%–30.50% of project selling price. The 12 items studied in the NEDO work, nine had quality costs in the range 4.50%–18.40% of project selling price. Looked at in this way, the quality costs of the vessel in this example look unacceptably high.

The largest quality costs lie in validation inspection (48.28% of total quality costs) and this is normal for fabrication. However, special procedures are the next highest cost at 20.61% of the total quality costs which is very high indeed and perhaps confirms a view, frequently expressed by fabricators in this industry, that offshore clients and contractors are unnecessarily demanding in terms of procedure qualification for special processes.

Repair and rectification costs are next at 13.45% and waiting time at 10.85% of quality related costs. The remaining items, control of bought-in items, quality engineering and auditing, amount to less than 7% of quality costs and less than 2% of project costs. They are almost negligible.

The cost of validation inspection is very high but this is not surprising. This is a highly inspection oriented business. The costs of auditing and special procedures were imposed by the client as part of contract requirements. The waiting time costs, in this instance, are arguably attributed to the client. Waiting time is, in fact, a complex item and very hard to cost. A very simple approach was adopted here of estimating waiting time in terms of lost man hours and costing this lost time at the man hour rate. This waiting time was attributed to waiting for third party inspection. It would be possible (if this were a very significant cost) to make a more detailed calculation and take into account the loss of useable capacity. Repair and rectification costs are a considerable proportion of quality costs and it seems, in this instance, that major repair was required. This is, without doubt, the fabricator's own responsibility but, of course, it can be argued that the client has some responsibility if he stipulates an excessive technical quality requirement in the purchase specification.

Analysis of QA Hours

In contrast to the previous example, the percentages for "quality" hours in column 4 of table 7.5 differ considerably from the percentages for quality costs in column 5 of the same table. This was because a considerable amount of quality related effort was subcontracted, so these percentage quality hours figures are not very useful as a measure of extent of all imput of quality related activities. However, in a small, relatively simple project, such as this, it is quite easy to estimate the man hours expended "in-house" by the fabricator and easy to obtain subcontracted costs directly from invoices rather than trying to investigate subcontractors' man hour inputs.

Comparison of the Two Examples

Client and manufacturer costs

The two examples described, although quite different in nature and looked at from quite different perspectives, were concerned with manufactured items for the offshore industry. Any such items are subject to manufacturing procedures and controls, special process controls, NDT and various forms of visual inspection and mechanical testing, together with the associated documentation and records. This is so for the fabrication and engineering industries in general, but, of course, in the offshore situation, the statutory requirements of certification can also play a significant part.

In the first example—the client controlled case—the vendors were not expected to carry any responsibility for quality assurance of the item. The client took full responsibility for this. So in addition to the normal costs paid for fabrication and installation, the client also paid for all the QA and QC that he expected to be carried out for the project. The costs to the client of doing this amounted to 13% of the project cost (table 7.2).

This assumes that:

1. The suppliers involved did not carry out any QA work themselves and did not bear any of the costs of QA.
2. The operator did not bear any costs of repair and rectification during the project. There was, in fact, no indication from available data that any significant repairs were needed. If they had been, the client

might well have needed to carry out extra testing or inspection, for which he would have borne the cost.

It is important to note that, for this first example, the costs of design, manufacture and installation were considered. In the second example, only the costs of design and manufacture were considered. In other words, the boundaries of each study were clearly defined and did not include the costs of any "quality problems" that might arise *after* installation and commissioning. This raises an interesting point. In the first example, the client bore all the costs of QA and took responsibility for it. In the event of future problems, he would carry the costs of loss of production *and* the responsibility for it. In the second example where the client carries the cost of loss of production, but where the fabricator bore the responsibility for, and cost of, QA, how well could the client recover any losses from this fabricator if there were in-service quality problems?

In this second example (the manufacture of a small pressure vessel) the fabricator bore all the direct costs of quality assurance and was responsible for implementing all QA activities. There was some additional QA in the form of extra inspection that the contractor required and also certifying inspection. The latter inspection was part of the certifying authority's procedures to fulfil UK statutory requirements.

Table 7.5 shows the quality related costs to be 25% of the project cost to the vendor. Most of these costs lie in validation inspection, but considerable costs lie in special procedures and repair and rectification. As with the first example, the highest costs also lie in validation inspection but do not vastly outweigh all other QA activities.

Special procedure and repair costs were also significant in this case and, as explained before, were *assumed* to be non-existent in the first example (likewise waiting time).

The assumptions made in this second example (the pressure vessel fabrication) were:

1. The nationally computed man hour rate applied to this fabricator was realistic for his situation.

2. "Waiting time" was realistically estimated in terms of man hours and could be reasonably costed by simple application of the company man hour rate.

For this fabricator, and the particular item studied, it was relatively easy to ascertain time and cost data by interviewing people and looking through past records. There was insufficient total man hour data (ie including subcontractor man hours), to establish (as in the first example) whether simple logging of man hour input would be adequate for a reasonable measure of quality costs. However, costs were reasonably straightforward to estimate, even on an "historical" basis. The client study was on a much larger scale and was obviously based on a much more complicated operation.

It needs to be stressed that these two examples, although sharing many features (in a QA sense), are not related in any way. For each situation, the items studied were parts of different projects and were for different clients. However, each item was for offshore use in the UK and each was, therefore, subject to certifying authority requirements. In the case of the small pressure vessel, this took the form of inspection by a certifying inspectorate. This small contract was only a part of a large project, and the inspection was a small part of the client's overall certifying requirements. In the case of the field modification example, the certifying input (£1 000) covered the whole project and there was no available data to break down exactly what activities were involved in each phase. Nevertheless, the cost of certification can be reasonably added as a pro-rata percentage to each phase.

In each example, subcontracted inspectors were involved but their roles were somewhat different. In the pipework project, the client operator appointed an inspection agency to carry out, and be totally responsible for, inspection. In the pressure vessel example, a major contractor, acting on behalf of a client operator, appointed a third party inspector to monitor inspection activities and records but the fabricator was still responsible for his own inspection commitments. The difference may be subtle but, nevertheless, very important in terms of ultimate accountability for inspection and quality assurance generally, as was implied earlier.

The quality requirements for each project were similar in terms of quality activities required (eg inspection, hydrotest, special approvals) and level of "engineering" quality. However, the overall quality related costs in each case are considerably different. On the basis of a very small sample, it is hard to say exactly why this may be. The client's quality costs, in the first example, were about 13% of project costs, and most of this was purely validation inspection. The fabricator's quality costs in the second example included special procedures and repair costs (negligible in the first example) and the total quality related costs were about 25% of project value. It could be argued that where clients require special procedures, the costs of these are negligible to clients, but considerable to fabricators. In the pipework contract example, we have no evidence as to whether the special procedures required were costly to the fabricators. Even if they were costly to the fabricators, in this instance, other QA costs were probably negligible, since the client carried responsiblity for this, anyway.

It is becoming clear that there are many complex, interacting factors at work in these two examples. However, to *summarise* in simple terms:

Fabricator quality related costs can be very high. This is borne out, not only from the offshore pressure vessel example quoted here, but from other similar studies (7.4). The high costs are caused by much inspection, and where special procedures, repair or delays are involved, these can make a significant contribution to quality costs. This was found to be the case where an offshore contractor, acting on behalf of an operator, required the fabricator to be responsible for quality assurance.

Client quality related costs are also very significant and in the situation quoted here, where the client carried total responsiblity for quality assurance, the costs mainly lay in validation inspection. These inspection costs, although considerable, were what the client would normally expect to apply—in other words, they were well within the client's control. This will be discussed more fully later on.

One could argue that the first example (the field modification project) should provide the model for project organisation if quality related costs are to be kept to a minimum. This model is only applicable, however, if the client has the engineering resources himself to select good suppliers and monitor their activities. This will not always be the case. If there is a complex contractual chain, as we have discussed in earlier chapters—particularly chapter 3, this will not be so and one can anticipate that total quality related costs will be higher.

CONCLUSIONS

The most important conclusion we should draw is that it is possible, certainly in the design and manufacturing stages without too much trouble, to identify and estimate quality related costs to the degree needed for identifying their relative magnitude—ie for management control purposes.

Figure 7.2, from work on quality costs in the process plant industry, shows that there is no real pattern of quality related costs in process plant manufacture, particularly for lower cost items. This work showed that individual contracts had individual quality cost patterns dependent on a variety of complex, interacting factors. This pattern is followed also in the offshore industry. The two examples discussed in detail in this chapter are examples of this. Each project was subject to similar quality activities such as NDT and hydrotest. The cost of "quality assuring" a project which was client controlled was of the order of 13% of project costs to that client. The cost of "quality assuring" a project which was fabricator controlled, on behalf of a client, was of the order of 25% of the contract price of the project.

Clearly, it would appear that the "cost of quality" is high and it seems disproportionately high to a fabricator. However, there are many factors, individual to each project, at work here and it would be dangerous to draw definitive conclusions on the basis of two examples. However, it is worth making some observations in relation to these examples. There was little evidence, in either case, of formal quality systems in operation. The client company studied did not expect any of its subcontractors to operate any formal system. The client was determined to be in control of all QA activity. The fabricating company studied did claim to operate a quality system but, with respect to the actual contract studied, there was considerable third party monitoring of quality and QA and a considerable requirement for special procedures (20% of quality costs, 5% of contract costs).

It could be argued that better management control on the part of the fabricator could have eliminated some of the quality costs (eg repair and rectification). It could also be argued that if the fabricator had operated an efficient and effective quality system (say to BS 5750), and client companies were confident of the fabricator's operation of this system, there would be much less need for third party involvement. If the fabricator also analysed his quality costs (say on a project by project basis), he could use these costs as a process analysis tool to pinpoint his problem areas (such as special processes) and hence use this as a budgeting or predictive tool for future tenders and contracts.

In the first example discussed, the client's quality related costs were negligible if measured against the increased oilfield production that the project would yield. The very high inspection costs (92% of quality costs, 13% of project costs) may be minimal to the client in terms of potential problems that might occur if this inspection had not taken place. It is interesting to note that special procedures approvals, which can be so costly to fabricators (refs 7.4, 7.7, 7.8), were a negligible client cost in this instance.

It would appear, in this case, that the client is in full control of his own costs, even though these are doubtless controlled to some degree by certifying authority requirements. There was insufficient scope in these examples to measure the impact of certification on QA costs, except that the fabricator "suffers" additional inspection and the client's QA manager would admit that his requirements tended to be conservative, "to be on the safe side". Both projects studied were good examples of the client problem highlighted by Hicks (ref 7.5): an inability or unwillingness to delegate (in the first example) and a requirement to monitor (in the second example).

One could now ask the question: if client companies demanded demonstration of acceptable quality systems by vendors and channelled their resources in assessing and monitoring these systems, would it be more cost effective for the client? One could further ask: if fabricating companies channelled their resources into operating, maintaining and reviewing their quality systems, would it be more cost effective for the fabricator? In the latter case, if the quality system cut down the need for clients to require special procedures, helped analyse the causes of need for repair and reduced waiting time, then it would be very cost effective for the fabricator.

The magnitude of the possible cost savings can be seen as worthy of consideration, at the very least. If the quality costs of any offshore project are of the order of 10-15% of total project cost then a saving of half of this is a lot of money.

However, a client company will always tend to take the view that when millions of pounds of potential oil and gas reserves are at stake,it makes little sense to "economise" on QA. QA in the offshore industry tends to be user driven and if client companies are reluctant to accept vendors' quality systems, then it is in vendors' interests to know their quality costs for budgeting and contractual purposes. This is perhaps the main, if unpalatable, message to the supplier organisations.

Nevertheless, it would be beneficial to clients and vendors alike to measure their quality costs, either on an annual or project related basis. Each should seek to formalise the gathering of quality cost data. It is not difficult to do and is good management accounting practice. Quality costs could be used to help in prediction and decision making and to highlight areas of inefficiency (as well as areas of strength, of course). Good management is being able, ultimately, to balance quality costs against profit. Efficient management of the former should lead to increase in the latter.

It is recommended not only that clients and vendors measure their own quality costs, but that some work should be done on measuring quality costs throughout a total project ie clients' costs, contractors' costs and all vendors' costs. This would help to indicate what proportion of particular activities' costs are carried by the various contractual parties. Such a study should also aim to measure the full impact of certifying authority requirements on quality costs.

REFERENCES

7.1 Plunkett, J. J. *Audits, Inspection and Costs in the Fabrication Sector of the Process Plant Industry*; MSc Thesis, UMIST, September 1982.

7.2 Feigenbaum, A. V. *Total Quality Control*, Third Edition, Chapter 7; McGraw-Hill, 1983.

7.3 Cox, B. *The Role of the management accountant in quality costing*; *Quality Assurance*, Vol 8, No 3, September 1982.

7.4 Moyes, E. M. and Rogerson, J. H. *The Reduction of Quality Related Costs in the Process Plant Industry*—Follow-on Study; NEDO Books, 1985.

7.5 BS 6143 Guide to the determination and use of quality related costs.

7.6 Hicks, J. G. Quality Assurance—at What Price?; *Metal Construction*, Vol 13, No 10, October 1981.

7.7 Nicholson, S. The case for national or industry based welding procedure pre-approval and pre-qualification schemes; *Metal Construction*, June 1984.

7.8 Salter, G. R and Bentley, K. P. *An analysis of the Effect of Welding Procedure Qualification Requirements on Production Costs in Process Plant and Offshore Structure Fabrication*; NEDO Books, 1985.